Lecture Notes in Networks and Systems

Volume 54

Series editor

Janusz Kacprzyk, Polish Academy of Sciences, Warsaw, Poland
e-mail: kacprzyk@ibspan.waw.pl

The series "Lecture Notes in Networks and Systems" publishes the latest developments in Networks and Systems—quickly, informally and with high quality. Original research reported in proceedings and post-proceedings represents the core of LNNS.

Volumes published in LNNS embrace all aspects and subfields of, as well as new challenges in, Networks and Systems.

The series contains proceedings and edited volumes in systems and networks, spanning the areas of Cyber-Physical Systems, Autonomous Systems, Sensor Networks, Control Systems, Energy Systems, Automotive Systems, Biological Systems, Vehicular Networking and Connected Vehicles, Aerospace Systems, Automation, Manufacturing, Smart Grids, Nonlinear Systems, Power Systems, Robotics, Social Systems, Economic Systems and other. Of particular value to both the contributors and the readership are the short publication timeframe and the world-wide distribution and exposure which enable both a wide and rapid dissemination of research output.

The series covers the theory, applications, and perspectives on the state of the art and future developments relevant to systems and networks, decision making, control, complex processes and related areas, as embedded in the fields of interdisciplinary and applied sciences, engineering, computer science, physics, economics, social, and life sciences, as well as the paradigms and methodologies behind them.

More information about this series at http://www.springer.com/series/15179

Nenad Mitrovic · Milos Milosevic
Goran Mladenovic
Editors

Experimental and Numerical Investigations in Materials Science and Engineering

Proceedings of the International Conference of Experimental and Numerical Investigations and New Technologies, CNNTech 2018

Springer

Editors
Nenad Mitrovic
Department for Process Engineering and Environmental Protection, Faculty of Mechanical Engineering
University of Belgrade
Belgrade, Serbia

Goran Mladenovic
Department for Production Engineering, Faculty of Mechanical Engineering
University of Belgrade
Belgrade, Serbia

Milos Milosevic
Innovation Center of Faculty of Mechanical Engineering
Belgrade, Serbia

ISSN 2367-3370 ISSN 2367-3389 (electronic)
Lecture Notes in Networks and Systems
ISBN 978-3-319-99619-6 ISBN 978-3-319-99620-2 (eBook)
https://doi.org/10.1007/978-3-319-99620-2

Library of Congress Control Number: 2018951894

This Springer imprint is published by the registered company Springer Nature Switzerland AG
The registered company address is: Gewerbestrasse 11, 6330 Cham, Switzerland

Preface

The book is a collection of high-quality peer-reviewed research papers presented at the International Conference of Experimental and Numerical Investigations and New Technologies (CNNTech 2018) held at Zlatibor, Serbia, from July 4 to 6, 2018. The conference is organized by the Innovation Center of the Faculty of Mechanical Engineering, Faculty of Mechanical Engineering at the University of Belgrade, and Center for Business Trainings. Over 30 delegates was attending the CNNTech 2018 —academicians, practitioners, and scientists from 11 countries—presenting and authoring 40 papers. The conference program included two keynote lectures with five invited lectures, four sessions (oral and poster), and two workshops. Seventeen selected full papers went through the double-blind reviewing process.

The main goal of the conference is to make positive atmosphere for the discussion on a wide variety of industrial, engineering, and scientific applications of the engineering techniques. Participation of a number of domestic and international authors, as well as the diversity of topics, has justified our efforts to organize this conference and contribute to exchange of knowledge, research results, and experience of industry experts, research institutions, and faculties which all share a common interest in the field in experimental and numerical investigations.

The CNNTech 2018 was focused on the following topics:

- Mechanical Engineering,
- Materials Science,
- Chemical and Process Engineering,
- Experimental Techniques,
- Numerical Methods,
- New Technologies.

We express our gratitude to all people involved in conference planning, preparation, and realization, especially to

- All authors, specially keynote speakers and invited speakers, who have contributed to the high scientific and professional level of the conference,
- All members of the Organizing Committee,

- All members of the International Scientific Committee for reviewing the papers and Chairing the Conference Sessions,
- Ministry of Education, Science and Technological development of Republic of Serbia for supporting of the Conference.

We wish to express a special gratitude to Ms. Dragana Perovic for her effort in preparing and managing the conference in the best way.

Organization

Scientific Committee

Miloš Milošević (Chairman)	University of Belgrade, Faculty of Mechanical Engineering, Serbia
Nenad Mitrović (Co-chairman)	University of Belgrade, Faculty of Mechanical Engineering, Serbia
Aleksandar Sedmak	University of Belgrade, Faculty of Mechanical Engineering, Serbia
Hloch Sergej	Technical University of Košice, Faculty of Manufacturing Technologies, Slovakia
Dražan Kozak	University of Osijek, Faculty of Mechanical Engineering in Slavonski Brod, Croatia
Nenad Gubeljak	University of Maribor, Faculty of Mechanical Engineering, Slovenia
Monka Peter	Technical University of Kosice, Faculty of Manufacturing Technologies, Slovakia
Snežana Kirin	University of Belgrade Innovation Center of faculty of Mechanical Engineering, Serbia
Ivan Samardžić	University of Osijek, Faculty of Mechanical Engineering in Slavonski Brod, Croatia
Martina Balać	University of Belgrade, Faculty of Mechanical Engineering, Serbia
Ludmila Mládková	University of Economics Prague, Czech Republic
Johanyák Zsolt Csaba	Athéné University, Faculty of Engineering and Computer Science, Hungary
Igor Svetel	University of Belgrade, Innovation centre of Faculty of Mechanical Engineering, Serbia
Aleksandra Mitrović	University of Belgrade, Faculty of Technology and Metallurgy, Serbia

Valentin Birdeanu	National R&D Institute for Welding and Material Testing - ISIM Timişoara, Romania
Danilo Nikolić	University of Montenegro, Faculty of Mechanical Engineering, Montenegro
Goran Mladenović	University of Belgrade, Faculty of Mechanical Engineering, Serbia
Darko Bajić	University of Montenegro, Faculty of Mechanical Engineering, Montenegro
Tasko Maneski	University of Belgrade, Faculty of Mechanical Engineering, Serbia
Luis Reis	IDMEC Instituto Superior Técnico, University of Lisbon, Portugal
Žarko Mišković	University of Belgrade, Faculty of Mechanical Engineering, Serbia
Tozan Hakan	Istanbul Medipol University, School of Engineering and Natural Sciences, Turkey
Nikola Momčilović	University of Belgrade, Faculty of Mechanical Engineering, Serbia
Traussnigg Udo	Institute for Electrical Machines and Drives University of Technology, Austria
Zoran Janjus	Faculty of Environmental Engineering, Bosnia and Herzegovina
Gordana Bakić	University of Belgrade, Faculty of Mechanical Engineering, Serbia
Katarina Čolić	University of Belgrade, Faculty of Mechanical Engineering, Serbia
Jovan Tanasković	University of Belgrade, Faculty of Mechanical Engineering, Serbia
Peter Horňak	Technical University of Košice, Faculty of Materials, Metallurgy and Recycling, Slovakia
Robert Hunady	Technical University of Kosice, Faculty of Mechanical Engineering, Slovakia
Martin Hagara	Technical University of Kosice, Faculty of Mechanical Engineering, Slovakia

Contents

Materials Science

Thermal and Mechanical Characteristics of Dual Cure Self-etching, Self-adhesive Resin Based Cement

Aleksandra Mitrovic[1(✉)], Nenad Mitrovic[2], Aleksandar Maslarevic[3], Vuk Adzic[2], Dejana Popovic[4], Milos Milosevic[3], and Dusan Antonovic[1]

[1] Faculty of Technology and Metallurgy, University of Belgrade, Karnegijeva 4, Belgrade, Serbia
aleksandramitrovic1926@gmail.com

[2] Faculty of Mechanical Engineering, University of Belgrade, Kraljice Marije 16, Belgrade, Serbia

[3] Innovation Center of Faculty of Mechanical Engineering, Kraljice Marije 16, Belgrade, Serbia

[4] Vinca Institute, University of Belgrade, Mike Petrovica Alasa 12-14, Belgrade, Serbia

Abstract. One of the main objectives in research and development of resin based cements (RBCs) is to enhance their clinical longevity and ease of use. In spite of the undeniable technological advances introduced in the last few decades, the polymerization shrinkage i.e. strain that accompanies the chain-growth polymerization of dimethacrylate monomers remains one of the major concerns for the clinical performance of composite restorations. Also, RBCs can produce a considerable amount of heat, due to the light energy from the curing lights and exothermic reaction of polymerization.

The purpose of this study was to determine the temperature changes during the photo-polymerization using thermocouples and to measure strain field of the self-etching, self-adhesive RBC, Maxcem Elite (Kerr, Orange, CA, USA) (ø5 × 1 mm - Group I and ø5 × 2 mm - Group II) using experimental technique, 3D Digital Image Correlation (DIC) method. Digital images were recorded immediately after photo-polymerization of the samples with a LED-curing unit for 20 s, according to manufacturer's recommendation. Vickers microhardness was determined after photo-polymerization and after 24 h. Temperature curves for both groups indicated similar patterns but the peak temperature of Group II was significantly higher compared to peak temperature of Group I. DIC showed that peripheral zone of the samples had the highest strain values in both groups. Group I indicated significantly higher values of hardness. All the results were material-dependent and probably correlated to the composition of each material, which is not fully disclosed by the manufacturers.

Keywords: Resin based cement · Temperature change · Thermocouples Strain · 3D Digital Image Correlation · Vickers microhardness

N. Mitrovic et al. (Eds.): CNNTech 2018, LNNS 54, pp. 3–15, 2019.
https://doi.org/10.1007/978-3-319-99620-2_1

1 Introduction

Resin based cements (RBCs) have extraordinary aesthetic shade-matching potential and due to the higher esthetical demands of patients for dental restorations, they have become very popular [1]. In an attempt to simplify procedures, a new group of RBCs, self-etching, self-adhesive resin cements (SARCs), have been developed. According to their manufacturers, these materials are self-adhesive, including acidic and hydrophilic monomers in their composition, which simultaneously demineralize and infiltrate enamel and dentin, providing strong bonding [2]. However, one of the shortcomings of RBCs is polymerization shrinkage that accompanies the chain-growth polymerization of dimethacrylate monomers and generates stress at the tooth and composite interface. The strain induced in RBCs during the polymerization can affect therapeutic failure due to de-cementation i.e. de-bonding. This strain depends on cement-layer thickness and presents the major source of shrinkage stress [3]. It has been shown that an increment thickness of 2 mm should not be exceeded for these materials with conventional chemistry [4]. Several methods were proposed to measure shrinkage strain such as dilatometry, the bonded disc method and strain gauges. Current methods include the Digital Image Correlation (DIC) method as an optical sensor for detecting shrinkage strain [5–7]. Heretofore, the hardness tests were widely used to examine RBCs and to analyze the efficiency of the light units, its wear resistance ability to keep stabile form and to have better insight into the degree of conversion (DC) during polymerization [8, 9]. Association of these tests with optical methods for strain analysis can reveal potential negative sides of the RBCs and improve different properties of currently available dental cements.

Another disadvantage of resin based cements is heat emission during polymerization [10]. The polymerization reaction of RBCs involves rupture of the C=C bonds of dimethacrylate monomers present in their polymeric matrixes and the conversion of intermolecular distances of 0.3–0.4 nm between polymer chains, maintained by Van der Waals attraction forces, into primary C–C covalent bonds [11]. Polymerization of these materials used for the fabrication of temporary restorations is associated with an exothermic reaction since the final formed enthalpy in the network is lower than that of the amount of the initial monomer. This temperature rise may cause thermal trauma to the pulp [12]. It has been proved that the pulp chamber is sensitive to physical, chemical, biological and thermal changes [13]. An increase of the intrapulpal temperature exceeding 42.58 °C can result in serious damage of the pulp tissue. Exothermic reactions of the composite resin and radiant heat from the light-curing unit contribute to heat production [10, 14]. Any curing unit that emits radiant energy in the blue area of the electromagnetic spectrum, i.e., between 400 and 500 nm, can be used to start the polymerization reaction of RBCs. Curing units with light-emitting diodes (LEDs) have been increasingly used lately. These light-curing units emit a narrow light spectrum with wavelengths close to the absorption peak of camphoroquinone (468 nm), i.e., the photosensitizer most used in RBCs [4].

Differential thermal analyses [15, 16], differential scanning calorimetry [17], infrared thermography [18, 19], thermistors [20, 21] and thermocouples [22–26] are some of methods that have been used to measure temperature increase during the polymerization of composite resin.

This study aims to determine the temperature changes during the photo-polymerization with LED lamp in a dual-cured self-etching, self-adhesive resin based cement, Maxcem Elite, using thermocouple, as well to measure strain and microhardness of investigated material.

2 Experimental Setup and Procedure

The research was carried out as an experimental study. The tested Maxcem Elite (Kerr, Orange, CA, USA) contains glycerol phosphate dimethacrylate (GPDM), co-monomers (mono-, di- and tri-functional methacrylate monomers, water, acetone, ethanol, barium, glass, fumed silica, and sodiumhexafluorosilicate) and ytterbiumfluoride mineral fillers.

The temperature measurement during the polymerization period was performed using a predefined procedure. The thermocouples used for temperature measurements were prepared using the OMEGA fine diameter bare wire (0.078 mm) K type (chromel-alumel) thermocouple, with special limits of error (±0.4%). The thermocouple bead was disk shaped to maximize temperature sensing surface area. The thermocouple was embedded into the center of the Maxcem Elite samples. The thermocouple voltages were acquired by HIOKI LR8431-20 high-speed multichannel data logger.

The teflon ring-type molds (ø 5 mm × 1 mm – Group I and ø 5 × 2 mm – Group II) were placed on a strip on the top of the quartz laboratory glass plate suspended on the laboratory stand. The LED curing light for the polymerization of both groups was placed just beneath the glass plate. The mold was filled with Maxcem Elite (Fig. 1). The samples were then cured for 20 s with a LED lamp (450–500 mW/cm^2, LEDition, Ivoclar-Vivadent, Schaan, Liechtenstein) according to manufacturer's recommendation. The temperature was recorded at 0.1 s intervals throughout the curing process and afterwards until it returned to the baseline level. The temperature of the top side of the glass plate, between the mold and the glass, was measured with a separate thermocouple in order to assess the heat input generated form the LED curing light. The extensive measurement showed that the heat input from the LED curing light was minimum and constant, contrary to the tested material polymerization temperature.

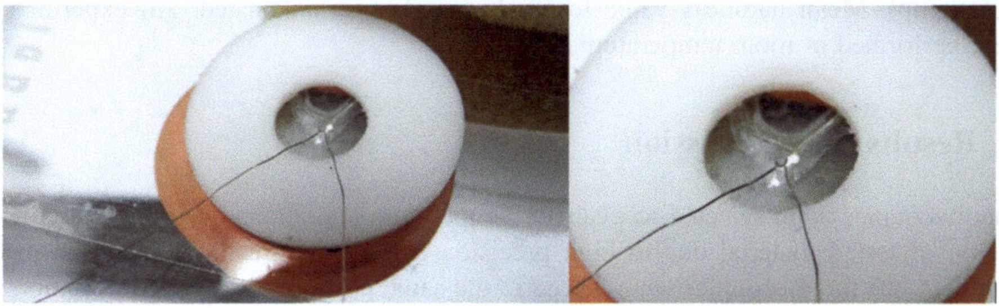

Fig. 1. Experimental setup for temperature measurement of Maxcem Elite during the polymerization.

Strain field was measured using 3D optical system Aramis 2M (GOM, Braunschweig, Germany) based on DIC method. The Aramis system consisted of two digital cameras with the resolution of 1600 × 1200 pixels and specialist software (Aramis v6.2.0). Prior to experiment, system calibration was performed using the calibration panel for corresponding measurement volume. This volume was chosen based on the dimensions of the measured area on sample surface. Two groups of the samples were prepared. First group (Group I) included three ø5 × 1 mm sized samples of Maxcem Elite (Kerr, Orange, CA, USA) and second group (Group II) included three ø5 × 2 mm samples of the same material. All samples were prepared by filling Teflon ring-type molds. The molds were placed on a strip on the top of the quartz laboratory glass plate suspended on the laboratory stand. The top surface of each sample was sprayed with fine black and white spray (Kenda Color Acrilico, Kenda Farben) to create a stochastic pattern with high contrast for image analysis. Digital images were recorded immediately after sample preparation (before polymerization, Stage 0) and immediately after photo-polymerization with LED lamp (after 20 s, Stage 1), in accordance with the manufacturer guidelines.

The images were then analyzed using software Aramisv6.2.0 to determine von Mises strain which represents criteria for the 3D deformation analysis in the x-, y- and z-axes of the photographed surface. Analysis of the strain fields was done using three sections (Sections 0, 1, 2). Circular section (Section 0) was positioned peripherally at the mold/material interface and two linear sections (Sections 1 and 2) were positioned orthogonally. Length of the Sections 1 and 2 corresponded to the sample diameter.

Microhardness measurements were performed with a digital device ECHOLAB type HTV 100 (Echo Research & Development S.p.a) that allows measurement of Vickers and Knoop micro-hardness. Device can be moved in two perpendicular directions, allowing distance measurement that ensures recommended distance (more than 2.5 d) between impressions. The force that can be applied to the device is 0.098–980.7 N (HV0.01–HV100). Measurements of the samples for both groups (N = 3) were carried out with a load of 0.49 N (HV0.05), so the ratio of sample thickness (1 mm and 2 mm) and mean-diagonal indentation was higher than 22 (min > 9) which meets the test requirements (sample thickness greater than 1.5 d). Loading time was 15 s. Vickers microhardness was measured for each group immediately after polymerization and after 24 h on the exposed surface (surface that was directly exposed to LED lamp). Mean hardness value for each sample was calculated. All experiments were performed at room temperature.

3 Results and Discussion

Thermocouples were selected to evaluate temperature alterations during the polymerization of tested material due to high precision and reliable readings. At each measurement, the thermocouples were placed in the same position to minimize variation in measurements that can be caused by changes in thermocouple position. The ambient temperature was constantly maintained because the temperature may affect the resultant net temperature rise for a given amount of energy dissipation, in accordance with Newton's law of cooling. Investigated material exhibited exothermic reaction during photo-polymerization. The temperature measured within the samples increased quite rapidly with the initiation of light curing. Group I reached maximum peak value of 46.2 °C within 15 s and Group II 54.2 °C within 19 s (Fig. 2). In accordance with the manufacturer guidelines, photo-polymerization lasted 20 s. The mean temperature increase for the Group II was higher than the mean temperature increase for the Group I. During the next 10 min, the temperature started to decrease to the values of 27.6 °C and 29.9 °C for the Group I and Group II, respectively. The curves reached plateau after 2 min. Temperature curves for both groups indicated similar patterns, but Group I showed less sharp increase in temperature than Group II. The peak temperature of Group II was significantly higher compared to peak temperature of Group I. This can be explained by the higher concentration of the accelerator due to the greater volume of the samples in Group II.

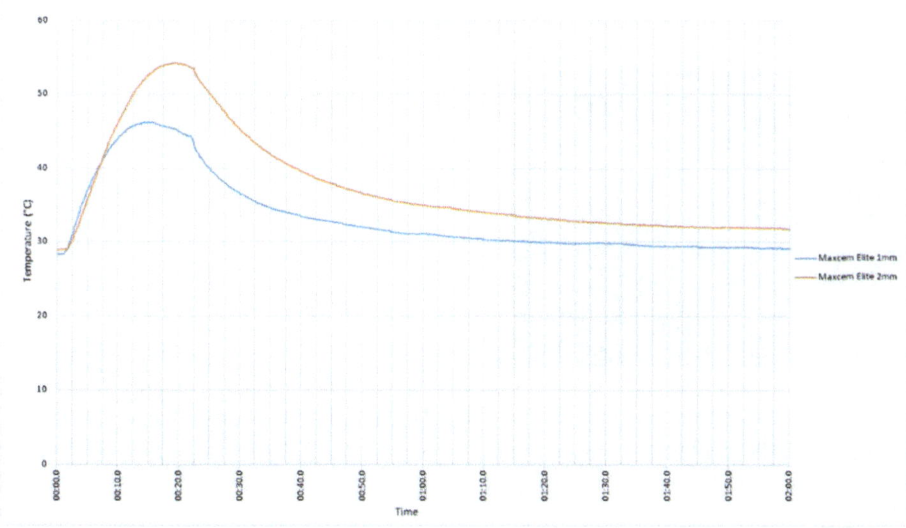

Fig. 2. Representative temperature curves for the Group I and Group II recorded during the photo-polymerization.

Both groups showed a gradual temperature reduction after the peak values due to the heat loss through dissipation to the surroundings exceeded the heat. The heat generated by the exothermic reaction mostly occurred in the initial phase of polymerization. Many studies have demonstrated a positive relationship between the light-curing unit intensity and temperature rise [19, 21, 23, 27]. Some in vitro studies reported that temperature rise depends on material used and photoactivation protocol i.e. depends, beside the light-curing unit type, on power density, exposure duration, the distance between composite surface and light guide tip end, composite shade and thickness of composite materials [28–30]. Many authors have quantified the amount of heat generated in resin-containing material during light curing. It was found that the temperature increase with LED lamps varies from 41 °C to 53 °C [31] which is consistent with our results. However, the measured temperatures recorded by Vallittu (6–40 °C) [32] were generally lower than measured temperature for Group I and Group II. This can be explained by the differences in the test conditions because the amount of exothermic heat during polymerization depends upon the amount of materials and the ambient temperature [33].

It is necessary to emphasize the limitations of in vitro studies (methodology). A significant amount of heat was generated in both groups of the investigated material, Maxcem Elite. Clinicians should be aware of the heat generated in RBCs during light-curing, which may be a potential source of pulpal injury.

The strain fields of representative samples of Group I (Fig. 3a and b) and Group II (Fig. 4a and b), after polymerization in Teflon molds, were shown.

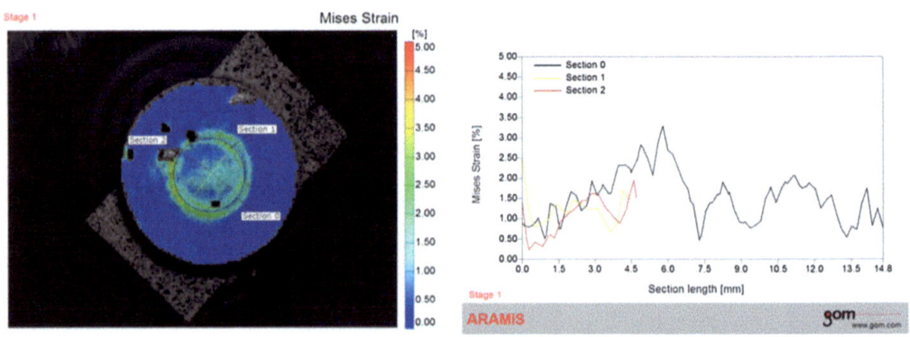

Fig. 3. (a) Shrinkage strain field at the Stage 1 of Group I; (b) von Mises strain as a function of distance for each section.

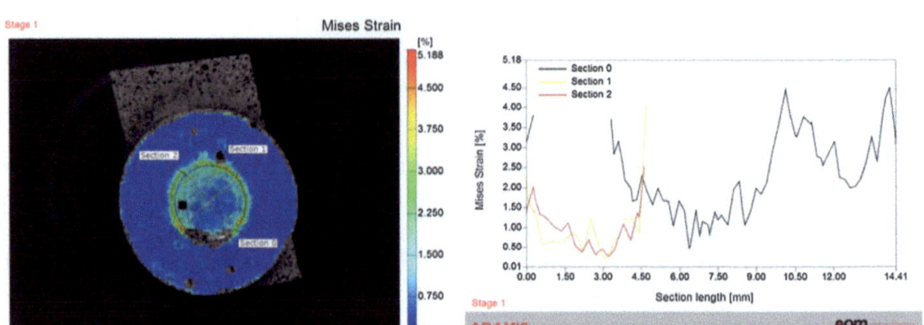

Fig. 4. (a) Shrinkage strain field at the Stage 1 of Group II; (b) von Mises strain as a function of distance for each section.

Section 0 (circular section) was placed at the mold/tested material interface. Section 1 and Section 2 represented linear sections that were orthogonally positioned. Group I and Group II showed a non-homogeneous strain field, especially in the center of the sample. Minor differences between mean strain values in the peripheral zone (Section 0) and central zone (Section 1 and Section 2) were presented for Group I samples (Table 1). The highest strain values (about 3.5%) were observed on the peripheral zone for Group I (Fig. 3a). For Group II, the highest strain values, about 4.0%, were noticed as well on the peripheral zone (Fig. 4a). However, major differences between mean strain values in the peripheral zone (Section 0) and central zone (Section 1 and Section 2) were observed for Group II. Obviously, peripheral zone was subjected to the highest strain values in both groups.

Table 1. Mean and standard deviation (SD) of von Mises strain (%) values.

Maxcem Elite ø5 × 1 mm (Group I)	Mean value of von Mises strain	Standard deviation
Section 0	2.206	0.704
Sections 1 and 2	1.578	0.709
Maxcem Elite ø5 × 2 mm (Group II)	Mean value of von Mises strain	Standard deviation
Section 0	2.084	0.843
Sections 1 and 2	0.897	0.674

It should be noted that the values in the central zone for Group I were significantly higher compared to Group II, since they had a similar value at the periphery. Possible explanation lies in fact that due to elongation, in absolute numbers, that is proportional to the material's initial dimension, thicker layers favor stress relief. Strain values reported in this study are consistent with dental composites, which range between 2% and 3% [34, 35].

Digital Image Correlation based on a two-camera 3D measurement system has been proven reliable for determining the strain i.e. polymerization shrinkage of RBCs [36–39]. In this study von Mises strain immediately after polymerization for both Maxcem Elite groups was measured. Unlike other methods for determining the dimensional stability of composite materials, the DIC method also enables the measurement of maximum strain value. In light-cured samples, the polymerization is much faster so the polymer matrix becomes semi-rigid in a few seconds. Therefore, immediately after light activation, the strain reached a high value, although the polymerization and crosslinking of the matrix were still on-going [40]. The study confirms previous findings that shrinkage behavior, including shrinkage magnitude of strain, is influenced by the curing mode in some degree [3, 4, 41, 42]. Particularly, this means that LED lamp may modify the overall strain pattern in terms of strain distribution. Previous studies [43, 44] have been conducted on the standardized sample size showing data for mean strain. These methods excluded peripheral section (Section 0) although the peripheral strain has to be considered when interpreting the overall strain. DIC method, in this study, detected maximal von Misses strain values and determined the zones of the maximal strain through presenting images of 3D full strain field. This study revealed additional information about curing mode dependent shrinkage patterns focusing on shrinkage and strain.

Vickers microhardness (HV) in both groups Maxcem Elite were measured immediately after the photo-polymerization and after 24 h. Mean microhardness values of the tested material are listed in Tables 2 and 3.

Table 2. Values of microhardness for Group I

Sample no.	Immediately after polymerization	Exposed surface
		HV 0.05
Sample 1	1	34.71
	2	38.52
	Average	36.62
Sample 2	1	35.22
	2	38.62
	Average	36.92
Sample 3	1	36.91
	2	34.62
	Average	35.77
Sample no.	After 24 h	Exposed surface
		HV 0.05
Sample 1	1	39.32
	2	36.81
	Average	38.07
Sample 2	1	37.46
	2	37.46
	Average	37.46
Sample 3	1	35.56
	2	35.48
	Average	35.52

Table 3. Values of microhardness for Group II

Sample no.	Immediately after polymerization	Exposed surface HV 0.05
Sample 1	1	25.13
	2	23.13
	Average	24.13
Sample 2	1	30.38
	2	31.01
	Average	30.70
Sample 3	1	26.08
	2	27.45
	Average	26.77
Sample no.	After 24 h	Exposed surface HV 0.05
Sample 1	1	28.73
	2	28.73
	Average	28.73
Sample 2	1	35.30
	2	30.94
	Average	33.12
Sample 3	1	29.77
	2	31.01
	Average	30.39

Considering the results for Group I, the values of microhardness immediately after polymerization and after 24 h were similar. Measured values of hardness were consistent as well for Group II. Group I indicated significantly higher values of hardness than Group II, which is in accordance with the results in the literature [5, 8, 22, 36]. Hardness is interpreted as a statistical magnitude that practically depends on the composition of the tested material structure, particularly in this case, for dual-phase structure like Maxcem Elite is. As the HV indenter is greater compared to the size of the fillers and the space in between filled with polymer matrix, the resulting HV value is a measurement of the filler-matrix system. The filler component is dominant factor compared to the softer polymer matrix [4]. HV indirectly considers the matrix network crosslinking. With 46% of filler volume fraction, Maxcem Elite is lower filled compared to other RBCs [45].

In dual-cure RBCs, there is a post-polymerization reaction within the polymer, which results in increased strain after a few minutes and higher values of hardness after 24 h. Lower content of fillers particles may have contributed to additional crosslinking or simple physical reorganization of polymer chains, which also leads to an increase in strain and hardness [46]. Proportional relationship between microhardness and polymerization shrinkage of the material has been shown by Li and co-workers [47]. This finding was in general agreement with other studies [45, 48], which revealed that the strain was related to the hardness. Previous studies investigated the microhardness of self-adhesive cements and found that the values greatly varied according to the brand [9], showing a very strong influence of the material [8].

4 Conclusions

Self-etch, self-adhesive resin cement, Maxcem Elite, was investigated. In the present study, the instructions of the manufacturer were followed with regard to the photo-polymerization time of the tested material. Results demonstrated that temperature increased inevitably after the initiation of the light curing for both groups. These increases were associated with damage to dental tissues. A significant amount of heat was generated in both groups. Clinicians should be aware that the heat generated in tested material during light-curing may be a potential source of pulpal injury. Group I and Group II showed a non-homogeneous strain field, especially in the center of the sample. However, peripheral zone was subjected to the highest strain values in both groups. Also, this fact highlights the advantage of the DIC method as a power tool for investigation in dentistry research fields. Group I indicated significantly higher values of hardness than Group II. The results were also material-dependent and correlated to the composition of the material. The role of each component on the final properties of the material has not been clarified yet. The complex formulation of tested Maxcem Elite is only partially disclosed by manufacturer, making it difficult to explain the strain differences through resin composition and inorganic content. It is important to bear in mind that all this study has been made in vitro and therefore, has some limitations. Further investigation will be conducted in order to better understand temperature and strain changes in the tested material.

Acknowledgement. The authors are grateful to Neodent (Belgrade, Serbia) for providing the material used in this study. This research was supported by Ministry of Education, Science and Technological Development of Republic of Serbia under Project TR35031.

References

1. Kitzmüller, K., Graf, A., Watts, D., Schedle, A.: Setting kinetics and shrinkage of self-adhesive resin cements depend on cure-mode and temperature. Dent. Mater. **27**, 544–551 (2011). https://doi.org/10.1016/j.dental.2011.02.004
2. Vrochari, A.D., Eliades, G., Hellwig, E., Wrbas, K.T.: Curing efficiency of four self-etching, self-adhesive resin cements. Dent. Mater. **25**, 1104–1108 (2009). https://doi.org/10.1016/j.dental.2009.02.015
3. Van Ende, A., Van de Casteele, E., Depypere, M., De Munck, J., Li, X., Maesc, F., Wevers, M., Van Meerbeek, B.: 3D volumetric displacement and strain analysis of composite polymerization. Dent. Mater. **31**, 453–461 (2015). https://doi.org/10.1016/j.dental.2015.01.018
4. Zorzina, J., Maiera, E., Harrea, S., Feyb, T., Belli, R., Lohbauer, U., Petschelt, A., Taschner, A.: Bulk-fill resin composites: polymerization properties and extended light curing. Dent. Mater. **31**, 293–301 (2015). https://doi.org/10.1016/j.dental.2014.12.010

5. Li, J., Fok, A.S., Satterthwaite, J., Watts, D.C.: Measurement of the full-field polymerization shrinkage and depth of cure of dental composites using digital image correlation. Dent. Mater. **25**, 582–588 (2009). https://doi.org/10.1016/j.dental.2008.11.001
6. Miletic, V., Manojlovic, D., Milosevic, M., Mitrovic, N., Stankovic, T.S., Maneski, T.: Analysis of local shrinkage patterns of self-adhering and flowable composites using 3D digital image correlation. Quintessence Int. J. Pract. Dentist. Engl. Ed. **42**, 797–804 (2011)
7. Fang Chen, T.Y., Sheng Huang, P., Fen, C.S.: Modeling dental composite shrinkage by digital image correlation and finite element methods. Opt. Lasers Eng. **61**, 23–30 (2014). https://doi.org/10.1016/j.optlaseng.2014.04.006
8. Cadenaro, M., Navarra, C.O., Antoniolli, F., Mazzoni, A., Di Lenarda, R., Rueggeberg, F. A., Breschi, L.: The effect of curing mode on extent of polymerization and microhardness of dual-cured, self-adhesive resin cements. Am. J. Dent. **23**, 14–18 (2010)
9. Giráldez, I., Ceballos, L., Garrido, M.A., Rodríguez, J.: Early hardness of self-adhesive resin cements cured under indirect resin composite restorations. Jo. Esthet. Restor. Dent. **23**, 116–124 (2011). https://doi.org/10.1111/j.1708-8240.2011.00408.x
10. Kim, R.J., Son, S.A., Hwang, J.Y., Lee, I.B., Seo, D.G.: Comparison of photopolymerization temperature increases in internal and external positions of composite and tooth cavities in real time: incremental fillings of microhybrid composite vs. bulk filling of bulk fill composite. J. Dent. **43**, 1093–1098 (2015). https://doi.org/10.1016/j.jdent.2015.07.003
11. da Silva, E.M., Penelas, A.G., Simão, M.S., Filho, J.D., Posku, L.T., Guimarães, J.G.: Influence of the degree of dentine mineralization on pulp chamber temperature increase during resin-based composite (RBC) light-activation. J. Dent. **38**, 336–342 (2010). https://doi.org/10.1016/j.jdent.2009.12.007
12. Michalakis, K., Pissiotis, A., Hirayama, H., Kang, K., Kafantaris, N.: Comparison of temperature increase in the pulp chamber during the polymerization of materials used for the direct fabrication of provisional restorations. J. Prosthet. Dent. **96**, 418–423 (2006). https://doi.org/10.1016/j.prosdent.2006.10.005
13. Ramoglu, S.I., Karamehmetoglu, H., Sari, T., Usumez, S.: Temperature rise caused in the pulp chamber under simulated intrapulpal microcirculation with different light-curing modes. Angle Orthod. **85**, 381–385 (2015). https://doi.org/10.2319/030814-164.1
14. Hannig, M., Bott, B.: In-vitro pulp chamber temperature rise during composite resin polymerization with various light-curing sources. Dent. Mater. **15**, 275–281 (1999). https://doi.org/10.1016/S0109-5641(99)00047-0
15. McCabe, J.F.: Cure performance of light-activated composites by differential thermal analysis (DTA). Dent. Mater. **1**, 231–234 (1985). https://doi.org/10.1016/S0109-5641(85)80048-8
16. Vaidyanathan, J., Vaidyanathan, T.K., Wang, Y., Viswanadhan, T.: Thermoanalytical characterization of visible light cure dental composites. J. Oral Rehabil. **19**, 49–64 (1992). https://doi.org/10.1111/j.1365-2842.1992.tb01590.x
17. Vaidyanathan, J., Vaidyanathan, T.K.: Computer-controlled differential scanning calorimetry of dental composites. IEEE Trans. Biomed. Eng. **38**, 319–325 (1991). https://doi.org/10.1109/10.133224
18. Al-Qudah, A.A., Mitchell, C.A., Biagioni, P.A., Hussey, D.L.: Thermographic investigation of contemporary resin-containing dental materials. J. Dent. **33**, 593–602 (2005). https://doi.org/10.1016/j.jdent.2005.01.010
19. Al-Qudah, A.A., Mitchell, C.A., Biagioni, P.A., Hussey, D.L.: Effect of composite shade, increment thickness and curing light on temperature rise during photocuring. J. Dent. **35**, 238–245 (2007). https://doi.org/10.1016/j.jdent.2006.07.012

20. Stewardson, D.A., Shortall, A.C., Harrington, E., Lumley, P.J.: Thermal changes and cure depths associated with a high intensity light activation unit. J. Dent. **32**, 643–651 (2004). https://doi.org/10.1016/j.jdent.2004.06.007
21. Shortall, A., El-Mahy, W., Stewardson, D., Addison, O., Palin, W.: Initial fracture resistance and curing temperature rise of ten contemporary resin-based composites with increasing radiant exposure. J. Dent. **41**, 455–463 (2013). https://doi.org/10.1016/j.jdent.2013.02.002
22. Hofmann, N., Hugo, B., Klaiber, B.: Effect of irradiation type (LED or QTH) on photo-activated composite shrinkage strain kinetics, temperature rise, and hardness. Eur. J. Oral Sci. **11**, 471–479 (2002). https://doi.org/10.1034/j.1600-0722.2002.21359.x
23. Leprince, J., Devaux, J., Mullier, T., Vreven, J., Leloup, G.: Pulpal-temperature rise and polymerization efficiency of LED curing lights. Oper. Dent. **35**, 220–230 (2010). https://doi.org/10.2341/09-203-l
24. Alnazzawi, A., Watts, D.C.: Simultaneous determination of polymerization shrinkage, exotherm and thermal expansion coefficient for dental resin-composites. Dent. Mater. **28**, 1240–1249 (2012). https://doi.org/10.1016/j.dental.2012.09.004
25. Mamourian, M., Esfahani, A.J., Ayani, B.M.: Experimental and scale up study of the flame spread over the PMMA sheets. Therm. Sci. **13**, 79–88 (2009). https://doi.org/10.2298/TSCI0901079M
26. Zhao, Z., Huang, H.M., Wang, Q., Ji, S.: Effects of pressure and temperature on thermal contact resistance between different materials. Therm. Sci. **19**, 1369–1372 (2015). https://doi.org/10.2298/tsci1504369z
27. Asmussen, E., Peutzfeldt, A.: Temperature rise induced by some light emitting diode and quartz-tungsten-halogen curing units. Eur. J. Oral Sci. **113**, 96–98 (2005). https://doi.org/10.1111/j.1600-0722.2004.00181.x
28. Hofmann, N., Markert, T., Hugo, B., Klaiber, B.: Effect of high intensity vs. soft-start halogen irradiation on light-cured resin-based composites. Part I. Temperature rise and polymerization shrinkage. Am. J. Dent. **16**, 421–430 (2003)
29. Daronch, M., Rueggeberg, F.A., Hall, G., De Goes, M.F.: Effect of composite temperature on in vitro intrapulpal temperature rise. Dent. Mater. **23**, 1283–1288 (2007). https://doi.org/10.1016/j.dental.2006.11.024
30. Knežević, A., Tarle, Z., Meniga, A., Sutalo, J., Pichler, G.: Influence of light intensity from different curing units upon composite temperature rise. J. Oral Rehabil. **32**, 362–367 (2005). https://doi.org/10.1111/j.1365-2842.2004.01418.x
31. Armellin, E., Bovesecchi, G., Coppa, P., Pasquantonio, G., Cerroni, L.: LED curing lights and temperature changes in different tooth sites. Biomed. Res. Int. **2016**, 1–10 (2016). https://doi.org/10.1155/2016/1894672
32. Vallittu, P.K.: Peak temperatures of some prosthetic acrylates on polymerization. J. Oral Rehabil. **23**, 776–781 (1996). https://doi.org/10.1046/j.1365-2842.1996.00430.x
33. Kim, S.H., Watts, D.C.: Exotherm behavior of the polymer-based provisional crown and fixed partial denture materials. Dent. Mater. **20**, 383–387 (2004). https://doi.org/10.1016/j.dental.2003.11.001
34. Braga, R.R., Ballester, R.Y., Ferracane, J.L.: Factors involved in the development of polymerization shrinkage stress in resin-composites: a systematic review. Dent. Mater. **21**, 962–970 (2005). https://doi.org/10.1016/j.dental.2005.04.018
35. Labella, R., Lambrechts, P., Van Meerbeek, B., Vanherle, G.: Polymerization shrinkage and elasticity of flowable composites and filled adhesives. Dent. Mater. **15**, 128–137 (1999). https://doi.org/10.1016/S0109-5641(99)00022-6
36. Miletic, V., Peric, D., Milosevic, M., Manojlovic, D., Mitrovic, N.: Local deformation fields and marginal integrity of sculptable bulk-fill, low-shrinkage and conventional composites. Dent. Mater. **32**, 1441–1451 (2016). https://doi.org/10.1016/j.dental.2016.09.011

37. Manojlovic, D., Dramićanin, M.D., Milosevic, M., Zeković, I., Cvijović-Alagić, I., Mitrovic, N., Miletic, V.: Effects of a low-shrinkage methacrylate monomer and monoacylphosphine oxide photoinitiator on curing efficiency and mechanical properties of experimental resin-based composites. Mater. Sci. Eng. C **58**, 487–494 (2016). https://doi.org/10.1016/j.msec.2015.08.054
38. Milosevic, M., Miletic, V., Mitrovic, N., Manojlovic, D., Savic Stankovic, T., Maneski, T.: Measurement of local deformation fields in dental composites using 3D optical system. Chem. Listy **105**, s751–s753 (2011)
39. Mitrović, A., Tanasić, V.I., Mitrović, N., Milošević, M., Tihaček-Šojić, L., Antonović, D.: Strain determination of self-adhesive resin cement using 3D Digital Image Correlation method. Arh. Celok. Lek, Srp (2018). https://doi.org/10.2298/sarh170530176m
40. Kaisarly, D., Gezawi, M.E.: Polymerization shrinkage assessment of dental resin composites: a literature review. Odontology **104**, 257–270 (2016). https://doi.org/10.1007/s10266-016-0264-3
41. Arrais, C.A., Rueggeberg, F.A., Waller, J.L., de Goes, M.F., Giannini, M.: Effect of curing mode on the polymerization characteristics of dual-cured resin cement systems. J. Dent. **36**, 418–426 (2008). https://doi.org/10.1016/j.jdent.2008.02.014
42. Kleverlaan, C.J., Feilzer, A.J.: Polymerization shrinkage and contraction stress of dental resin composites. Dent. Mater. **21**, 1150–1157 (2005). https://doi.org/10.1016/j.dental.2005.02.004
43. Spinell, T., Schedle, A., Watts, D.C.: Polymerization shrinkage kinetics of dimethacrylate resin-cements. Dent. Mater. **25**, 1058–1066 (2009). https://doi.org/10.1016/j.dental.2009.04.008
44. Martinsen, M., El-Hajjar, R.F., Berzins, D.W.: 3D full field strain analysis of polymerization shrinkage in a dental composite. Dent. Mater. **29**, e161–e167 (2013). https://doi.org/10.1016/j.dental.2013.04.019
45. Frassetto, A., Navarra, C.O., Marchesi, G., Turco, G., Di Lenarda, R., Breschi, L., Ferracane, J.L., Cadenaro, M.: Kinetics of polymerization and contraction stress development in self-adhesive resin cements. Dent. Mater. **28**, 1032–1039 (2012). https://doi.org/10.1016/j.dental.2012.06.003
46. Truffier-Boutry, D., Demoustier-Champagne, S., Devaux, J., Biebuyck, J.J., Mestdagh, M., Larbanois, P., Leloup, G.: A physico-chemical explanation of the post-polymerization shrinkage in dental resins. Dent. Mater. **22**, 405–412 (2006). https://doi.org/10.1016/j.dental.2005.04.030
47. Li, J., Li, H., Fok, A.S.L., Watts, D.C.: Multiple correlations of material parameters of light-cured dental composites. Dent. Mater. **25**, 829–836 (2009). https://doi.org/10.1016/j.dental.2009.03.011
48. Meira, J.B.C., Braga, R.R., Ballester, R.Y., Tanaka, C.B., Versluis, A.: Understanding contradictory data in contraction stress tests. J. Dent. Res. **90**, 365–370 (2011). https://doi.org/10.1177/0022034510388039

TGA-DSC-MS Analysis of Pyrolysis Process of Various Biomasses with Isoconversional (Model-Free) Kinetics

Nebojsa Manic[1(✉)], Bojan Jankovic[2], Dragoslava Stojiljkovic[1], and Vladimir Jovanovic[1]

[1] Faculty of Mechanical Engineering, University of Belgrade, Kraljice Marije 16, Belgrade, Serbia
nmanic@mas.bg.ac.rs
[2] Faculty of Physical Chemistry, University of Belgrade, Studentski trg 12-16, Belgrade, Serbia

Abstract. Slow pyrolysis characteristics of different biomasses (hazelnut shell (HS), sawdust (Beech), and sawdust chemically treated (SDCT)) were investigated by simultaneous thermal analysis (STA), coupled with mass spectrometry (MS). Thermal decomposition of these samples was divided into three stages corresponding to removal of water, devolatilization, and formation of bio-char. It was found that differences in thermal behavior of the samples are due to differences in their composition. Mass spectrometry results showed that H_2, CH_4, H_2O, CO_2 (C_3H_8), CO, and C_2H_6 were the main gaseous products released during pyrolysis. It was shown that HS could be a good fuel, since during its pyrolysis at high temperature, more gaseous products are released compared to other systems. Isoconversional (model-free) method was used in order to determine variation magnitudes of effective activation energy (E_a) values on conversion fraction (α) during pyrolysis. It was found that identified variations of E_a with α arise from the different chemical structures among cellulose, hemicelluloses and lignin in tested samples that may affect on their effective activation energies.

Keywords: Biomass · Volatilization · Model-free · Pyrolysis mechanisms Devolatilization index

1 Introduction

The energy from the Sun is stored in biomass through the process of photosynthesis. The photosynthesis starts with the capture of light by photosynthetic accessory pigments and conversion into electrical energy by the chlorophyll pigments of reaction centers. In a next stage, not photochemical, although light is required to activate certain enzymes, the stored chemical energy is used to reduce carbon dioxide (CO_2) and resulting synthesis of carbohydrates.

Biomass is a carbon dioxide neutral option for energy generation, and potential for the conversion of biomass into energy is rapidly increasing [1, 2]. According to Maniatis [3], the energy from biomass based on a short rotation forestry and other

N. Mitrovic et al. (Eds.): CNNTech 2018, LNNS 54, pp. 16–33, 2019.
https://doi.org/10.1007/978-3-319-99620-2_2

energy crops can significantly contribute the objectives of the Kyoto Agreement in reducing the greenhouse gas emissions and the problems related to climate change. Nevertheless, a number of biomass technologies are available for converting the biomass to energy. These processes can change raw biomass into a variety of gaseous, liquid, or solid materials that can then be used for energy generation. This conversion can be implemented in three ways: thermo-chemical (high temperature decomposition of biomass), bio-chemical (biomass decomposition through microorganism or enzymatic processes) and chemical (oils from biomass can be chemically converted into a liquid fuel) [4, 5].

In recent years, there have been many developments in science and technology related to thermo-chemical biomass conversions. Incineration, gasification and pyrolysis conversion are among established and best available thermo-chemical technologies. These thermal processes provide an efficient, environmentally acceptable and the cost-effective method of providing a sustainable energy source.

Pyrolysis processes are conducted in an inert atmosphere for production of solids (bio-char), liquids (bio-oils), or gases with high heating value. Their relative fractions depend on the operating conditions (the extreme situations are usually referred to slow pyrolysis ($\sim$400 °C using the slow heating rates characterized by high vapor residence time (5–30 min)) and fast pyrolysis (which mainly maximize the production of bio-oil, where the biomass is quickly heated to a high temperature at very high heating rates: 1000–10.000 °C s^{-1})).

Bio-char represents the solid product obtained from pyrolysis. High values of carbon content and heating value in bio-char made it comparable to coal and coke used in a blast furnace [6]. Process parameters of pyrolysis such as temperature, heating rate, residence time, pressure, and particle size are crucial, in particular to achieve high quality of bio-char. Among the parameters, temperature is the most influenced parameter on bio-char quality and yield. Increase of pyrolysis temperature leads to better quality of bio-char, i.e. higher carbon content, lower volatile content, lower ash content, and higher heating value. On the other side, at high pyrolysis temperature bio-char yield is lower, which corresponds to poor economic performance of pyrolysis processes [7].

Through pyrolysis process, it is possible to obtain a higher value-added liquid fuel with, i.e. bio-oil. This fuel is manageable in combustion processes [8] and generates lower toxic emissions compared to those of petroleum fuels. Bio-oil is a biofuel obtained by condensation of vapors arising from pyrolysis of biomass. However, bio-fuel is liquid product of the fast pyrolysis of biomass. Many researchers have been engaged in its application and the further treatment. One of important products of pyrolysis are gaseous fuels (hydrogen and syngas) [9] where other valuable gases, such as carbon monoxide (CO) can also be generated by pyrolysis.

The main goal of this work is the characterization of pyrolysis process of different types of biomass, using the thermogravimetric analysis (TGA)/derivative thermo-gravimetry (DTG) – differential scanning calorimetry (DSC) coupled with mass spectrometry (MS) detection analysis. Obtained experimental data were further used for kinetic investigation of pyrolysis, through application of isoconversional (model-free) methods [10–13]. Since actual reaction mechanism of biomass pyrolysis is extremely complex due to highly heterogeneous process [14] and diversity of its constituents (hemicelluloses, cellulose and lignin, which are distinguished by their own thermal decomposition paths), the kinetic information is necessary for optimization of entire pyrolytic process, in targeting the volatilization kinetics.

2 Materials and Methods

2.1 Materials and Sampling Procedure

For simultaneous thermal analysis (STA), three different biomass samples were used. One of tested samples is agricultural biomass sample - hazelnut shells and the remaining two samples represent wood processed products: sawdust and factory-formed wood product as chemically treated sawdust (treated with organic adhesive – MDF (medium-density fiberboard) sawdust (urea-formaldehyde (UF) resin)).

Hazelnut shells are collected from suburban region of Serbia, near Capital city, Belgrade. After removing hazelnuts' inner part, shells were collected in bags and left for 8 months before they were brought to laboratory. Sawdust was collected from the process of cutting Beech log wood for household heating, while chemically treated sawdust (MDF) was obtained from local carpenter workshop.

According to standard EN ISO 14780 [15] for solid biomass fuel sample, the preparation of each biomass sample was prepared for experimental tests. Sample was removed from the transportation packing, and then placed inside the oven at 105 °C for 2 h, in order to define moisture left on the inner surfaces of the packing. This amount of moisture was included in the calculation of the total moisture content in the sample according to standard EN ISO 18134 [16]. Further, regarding to pre-drying process of the sample, which is necessary to remove the residual moisture during preparation, sample was placed on a plate and left to reach moisture equilibrium with laboratory atmosphere conditions for 24 h. After that period of time, sample was reduced in two steps using cutting mill, after which it was sieved through the 1 mm (18 Mesh sizes) sieve.

Finally, the obtained undersize was declared as a general analysis sample and divided into three test portions for further testing. First portion of prepared sample was used for determination of ultimate analysis (carbon, hydrogen and nitrogen content) according to standard EN ISO 16948 [17]. The results of proximate analysis (total moisture, ash, volatile matter and char content) were obtained from the second portion of the prepared sample, according to standard procedures EN ISO 17225-1 [18]. The higher heating value (HHV) and lower heating value (LHV) were determined using calorimeter laboratory equipment with an oxygen bomb (IKA C200, IKA® Works, Inc., Wilmington, USA), according to the standard EN ISO 18125 [19].

2.2 Simultaneous Thermal Analysis (STA) Measurements

The STA (TGA-DSC) measurements were performed on the third portion of the prepared biomass sample. NETZSCH STA 445 F5 Jupiter system (Erich NETZSCH GmbH & Co. Holding KG, Germany) was used for STA experimental tests for all examined samples. Inert atmosphere was provided to maintain the pyrolysis process using the high purity nitrogen (Class 4.6) as a carrier gas. At the same time, nitrogen was used as a protective gas in order to keep the high sensitive internal balance (0.1 µg). Both carrier and protective gas flow were set to the φ = 50 mL min^{-1}. The weight measurements were carried out using internal balance which provided the following results: Hazelnut shell - 5.35 ± 0.10 mg, Sawdust - 5.70 ± 0.50 mg, and MDF sawdust - 5.70 ± 0.50 mg, respectively. Crucibles that were used for tests are made of alumina, and during each measurement they were filled approximately up to the half. Crucible's lid was placed on the top, so the optimum heat transfer could be realized. Each sample was tested using three different heating rates (β = 5, 10 and 20 °C min^{-1}). Using these heating rates, the samples were heated from room temperature up to 600 °C. During all measurements, the sample temperature controller (STC) was turned off, so the set temperature (600 °C) is referred to furnace temperature. Temperatures presented on diagrams are sample temperatures that, due to construction of furnace, never reached the set temperature. This fact resulted in better temperature curve linearity, than it would be if STC was turned on. The STA 445 F5 Jupiter runs under the versatile Proteus® software and includes all operating tools to obtain a reliable measurement and evaluate the resulting data, or even carry out complicated analyses.

2.3 Mass Spectrometry (MS) Analysis

Determination of evolved gases from performed TGA (STA) was carried out continuously using the quadrupole mass spectrometer NETZSCH QMS 403 D Aëolos (QMS) (Erich NETZSCH GmbH & Co. Holding KG, Germany). TGA/STA – QMS coupling was done using transfer line with quartz capillary tube with diameter of 75 µm. The whole transfer line was heated up to 230 °C in order to avoid condensation of evolving volatile compounds. The QMS was operated under vacuum up to 10^{-7} bar, providing conditions necessary to detect gas components using their ions intensity according to their respective mass to charge ratios (m/z).

Evolved gas composition was monitored and analyzed through bar-graph cycles, scanning mass units in the range from 1 to 80. Cycles were set to speed of 0.2 s per mass unit and carried out using stair mode. The excitation energy in the QMS was set up at 1200 eV with the resolution of 50 units.

From selected range, the focus was on specific ions that correspond to gases evolved in pyrolysis process. Accordingly, the molecules with atomic mass units (amu) of 2, 16, 18, 28, 30, 44, 58 and 72 which correspond to H_2, CH_4, H_2O, CO, C_2H_6, C_3H_8 (CO_2), C_4H_{10} and C_5H_{12} respectively, were analyzed, and the intensity peak areas obtained for each compound were compared.

2.4 Theoretical Background

Thermogravimetry (TG) analysis is a common and generally repeatable method for conducting biomass pyrolysis experiments for purpose of extracting apparent kinetic parameters [20]. TG records thermal events that correspond to chemical or physical changes in the sample by measuring gain or loss in the sample mass as a function of temperature or time. A dimensionless extent of reaction term known as conversion fraction, α, was calculated from experimental data as follows $\alpha = 1 - (m_t - m_\infty)/(m_o - m_\infty)$, where m_t is the mass at time t, m_o is the measured mass at the beginning of TGA experiment, and m_∞ is the measured mass at the end of TGA experiment. In general, the rate of biomass decomposition has been expressed by the rate-law equation in the isothermal conditions as:

$$\frac{d\alpha}{dt} = k \cdot (1 - \alpha)^n \quad (1)$$

where $(1 - \alpha)$ is the unreacted biomass fraction at time t; n is the reaction order, and k is the rate constant given by the Arrhenius equation $k = A \cdot \exp(-E_a/RT)$ where A is the pre-exponential factor, E_a is the effective (apparent) activation energy for a given process, R is the universal gas constant, and T is the absolute temperature. To extract the kinetic parameters obtained from non-isothermal (dynamic) TGA data, a linear heating rate (β) in a form $\beta = dT/dt$ is typically used, so that the temperature-dependent rate-law equation for decomposition of biomass in non-isothermal conditions becomes:

$$\beta \cdot \left(\frac{d\alpha}{dT}\right) \equiv \left(\frac{d\alpha}{dt}\right) = A \cdot \exp\left(-\frac{E_a}{RT}\right) \cdot (1 - \alpha)^n \quad (2)$$

Many researches have applied integral approach to extract kinetic parameters of materials from non-isothermal thermo-analytical (TA) data at the different heating rates [21, 22]. Other widely used integral methods include those developed by Kissinger [23], as well as by Akahira and Sunose [24] known as KAS method. The rate equation as shown in Eq. (2) has also been used in its differential form to obtain kinetic parameters described by Friedman [25, 26], known as the Friedman's (FR) method. Models established by these researchers offer convenience of obtaining kinetic information without prior knowledge of exact reaction mechanism. In addition, the assumption that the conversion fraction, α, for a reaction is constant (isoconversional) simplifies the calculation of the kinetic triplet (A, E_a and $f(\alpha)$; $f(\alpha)$ is the conversion function which depends on the mechanism of the thermal decomposition) through the linearization techniques. Both of aforementioned methods (differential - FR and integral - KAS) were used in this work to perform dependence $E_a = E_a(\alpha)$. Kinetic analysis based on model-free methods allows kinetic parameters to be evaluated for different constant extents of conversion without evaluating any particular form of reaction model. Many solid-state reactions follow complex mechanisms including multiple series and parallel stages with different E_a's. Ability of model-free methods to show the complexity is a crucial step toward ability to explain mechanistic conclusions from kinetic data.

3 Results and Discussion

3.1 Results of Proximate and Ultimate Analysis and Bio-char Yields

Table 1 shows the results of proximate and ultimate analysis of tested biomasses (hazelnut shell, sawdust and chemically treated sawdust (MDF sawdust), respectively).

Table 1. The proximate and ultimate analysis of raw biomass samples

Proximate analysis (wt%)		Ultimate analysis[b]	
Hazelnut shell			
Moisture	9.27	C	50.12
Volatile matter	68.8	H	6.66
Fixed carbon	20.7	Oc	40.53
Ash	1.67	N	0.86
HHV (MJ kg^{-1})	18.0	S	0.00
LHVa (MJ kg^{-1})	16.8		
Sawdust (Beech)			
Moisture	7.16	C	49.46
Volatile matter	74.4	H	6.82
Fixed carbon	17.2	O[c]	41.88
Ash	1.68	N	0.03
HHV (MJ kg^{-1})	17.8	S	0.00
LHVa (MJ kg^{-1})	16.1		
Chemically treated sawdust (Beech)			
Moisture	5.40	C	48.63
Volatile matter	79.2	H	6.55
Fixed carbon	14.0	O[c]	42.20
Ash	0.58	N	2.01
HHV (MJ kg^{-1})	18.5	S	0.00
LHV[a] (MJ kg^{-1})	16.2		

[a]Calculated according to [19]
[b]On a dry basis
[c]By the difference

It can be seen from Table 1 that all biomass samples are characterized by high volatiles matter, ranging from 68.08–79.62%, which makes them desirable for a good regulation of combustion or gasification processes. The raw materials have moisture content in a range 5.40–9.27% which can be attributed to the open air conditions. It can be observed that MDF sawdust is characterized by the highest value of volatile matter in respect to other samples, which means it is characterized by the highest conversion, in comparison to biomass sample with the highest fixed carbon (hazelnut shell, Table 1). From a theoretical point of view, sawdust and MDF sawdust can be suitable for good production of bio-oil, while the hazelnut shell would be the best biomass sample for the production of bio-char (Table 1). All tested samples have the HHV values that belong to heating values of biomass derived bio-chars (11.83–44.20 MJ kg^{-1}) [27, 28].

In addition, the concentration of C and H is highest for hazelnut shell and sawdust (Table 1). Almost negligible content of N has been identified for all biomass samples, except for MDF sawdust (Table 1), which means that the potential toxic NOx emission during the studied process is almost eliminated. However, all samples show the presence of oxygen more than 30% which means that they can be essential in deoxygenation.

High carbon and low oxygen in biomass as compared to coal (73.1% of carbon and 8.7% of oxygen) are favorable for combustion applications [29, 30], while the higher proportion of carbon (relative to hydrogen and oxygen content) (hazelnut shells in Table 1) increases the energy content of a fuel because the energy contained in carbon-carbon bonds is greater than that of carbon-oxygen and carbon-hydrogen bonds [31].

The bio-char yield for considered samples was calculated from STA experiments using a relation: Bio-char yield (%) = $[(m_f^*)/(m_i)] \times 100$, where m_f^* is the mass at desired temperature and time, while mi is the mass at time t = 0 (min). Table 2 lists the values of bio-char yields of biomass samples in pyrolysis process obtained at different operating temperatures and residence times at $\beta = 10$ °C min^{-1}.

Table 2. Bio-char yields at different temperatures and residence times for $\beta = 10$ °C min^{-1}

Hazelnut shell				
T (°C)	250	350	450	550
t (min)	24.47	33.75	43.55	53.26
Yield (%)	88.00	52.98	36.24	31.66
Sawdust (Beech)				
T (°C)	250	350	450	550
t (min)	24.80	34.08	43.68	53.77
Yield (%)	88.59	50.07	28.76	25.09
Chemically treated sawdust - MDF (Beech)				
T (°C)	250	350	450	550
t (min)	24.63	34.17	43.70	53.49
Yield (%)	91.55	49.85	26.08	22.82

It can be seen from Table 2 that for all samples, the bio-char yields are negatively correlated with increasing pyrolysis severity (increased operating temperature and longer residence times). Among tested samples, sawdust and MDF sawdust have the lowest bio-char yield for $T \geq 450$ °C along residence time $t > 43.50$ min. At high operating temperatures, probably, there is a significant distinction in composition of cell wall components and extractives in wood by-products in comparison with agricultural product. From presented results, the hazelnut shell has the highest bio-char yield (except at the lowest operating temperature). Generally, the bio-char yields (over considered *T*–*t* – ranges) in this study amount 49.85% which is more than the yield received for algal biomass [32]. On the other hand, the obtained results of bio-char yields fall in the range for bio-char yield ranges under slow pyrolysis of biomass feedstock (2–60%) [33].

3.2 TGA-DTG-DSC Curves of Pyrolysis Processes of Various Biomass Samples

Figures 1, 2 and 3 show experimental TG (Thermogravimetric) – DTG (Derivative thermogravimetric) – DSC (Differential scanning calorimetry) curves for thermal decomposition of hazelnut shell (a), sawdust (b) and chemically treated sawdust (MDF) (c) under an atmosphere of nitrogen.

It can be observed that with an increasing of heating rate, TGA curves are shifted to the right, i.e. that the mass losses occur at increasing temperatures. This behavior has been described by several researchers and can be explained using different arguments [34–36]. Some authors argue that this behavior is due to changes in reaction mechanism caused by increased heating rate. Furthermore, poor heat transmission to the sample in the oven may cause increasing differences with increasing heating rate between the nominal and real temperature of the sample. It could also be due to different rates of heat dissipation or absorption of the reaction at the different heating rate. However, the observed shift can be simply explained by mathematical form of kinetic laws, which can provide a shift of the curves at higher temperatures with increasing reaction rate with the same kinetic constants.

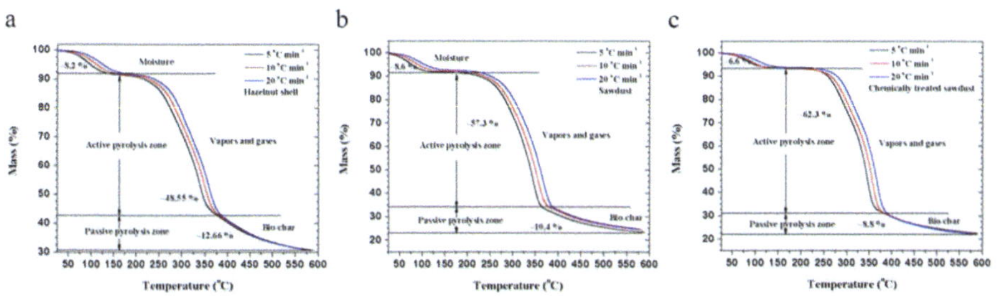

Fig. 1. TGA curves of (a) hazelnut shell, (b) sawdust (Beech), and (c) chemically treated sawdust (MDF) pyrolysis process, at the different heating rates (5, 10 and 20 °C min^{-1})

All TGA curves show three main stages which are typical for biomass pyrolysis process. The first stage corresponds to moisture release (water evaporation) (pre-pyrolysis stage), which is approximately performed in the temperature range 25–175 °C. The remaining two stages are attributed to active and passive pyrolysis, respectively. The active pyrolysis stage is characterized by the largest mass loss where various vapors and gases are released as a result of decomposition processes of hemicelluloses and cellulose which takes place in this stage, within temperature ranges 175–380 °C and 250–380 °C, respectively. On the other hand, the lignin is decomposed in both stages: active and passive pyrolysis stages in the temperature range 175–600 °C, without characteristic peaks (Fig. 2). However, the various biomasses show different heat effects in respect to the number of endothermic and exothermic peaks at the fixed heating rate (Fig. 3). The depth of endothermic peak related to pre-pyrolysis stage varies from sample to sample, which is in correlation with moisture content in samples. The two endothermic peaks at the DSC curves that can be attributed to decompositions of hemicelluloses and cellulose do not appear in all samples, and this is refers sawdust and MDF sawdust (Fig. 3). One of processes is certainly "masked" given that probably there is an advantage of overlapping decomposition reactions, related to one of the pseudo-components. DSC curve for the hazelnut shell is only distinguished from other DSC curves. Certain variations in the heat effects can be significantly stimulated by concentration of three pseudo-components (hemicelluloses, cellulose, and lignin) present in samples, which also affects the development of

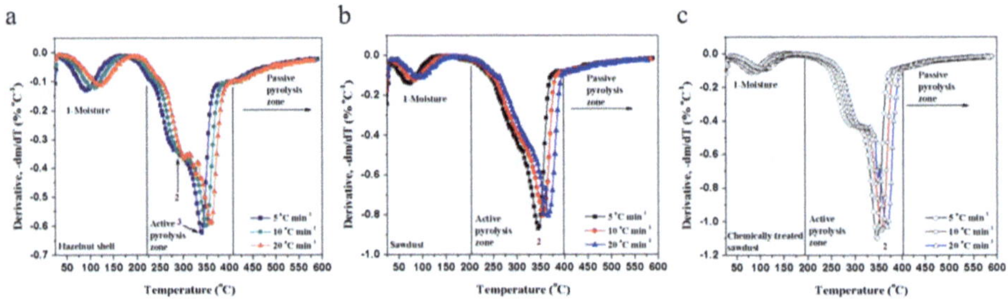

Fig. 2. DTG curves of (a) hazelnut shell, (b) sawdust (Beech), and (c) chemically treated sawdust (MDF) pyrolysis process, at the different heating rates (5, 10 and 20 °C min^{-1})

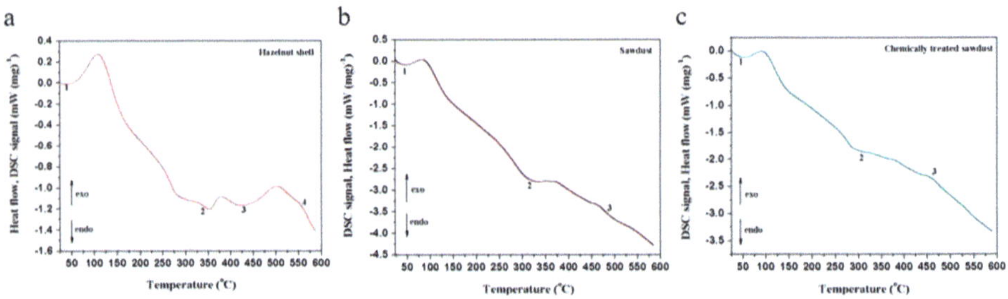

Fig. 3. DSC curves of (a) hazelnut shell, (b) sawdust (Beech), and (c) chemically treated sawdust (MDF) pyrolysis process, at β = 10 °C min^{-1}

decomposition paths. For sawdust and MDF sawdust, the slight exothermic effect in passive pyrolysis which is attributed to bio-char forming was designated by "3", while for hazelnut shells this is indicated by "4". Strength of endothermic peak caused by the volatilization of moisture can be enhanced by the higher gas flow, while with an increase of pyrolysis temperature, the heat of exothermic reaction during pyrolysis can increased gradually. Considering all DTG curves (Fig. 2), the long tails in a passive pyrolysis stage can be attributed to decomposition of lignin at the temperatures above 400 °C, where pyrolysis proceeds with a slower mass loss rate.

It can be concluded that at high heating rates, separate DTG peaks did not arise because some of them were decomposed simultaneously and several adjacent peaks were united to form overlapped boarder and higher peaks. This is consequence of heat and mass transfer limitations. In this respect, with an increase in heating rate, the temperature in furnace space can be a little higher as temperature of a particle and the rate of decomposition are higher than the release of volatiles. Differences in molecular structures and chemical nature of three pseudo-components account for dissimilar behaviors. It is obvious that variations in appearance of DTG curves are primary related to variation in percentage of hemicelluloses presented in biomass samples. Namely, higher lignin content may leads to a slower decomposition with a more energy needed, whereas higher cellulose and hemicelluloses content decomposes faster and produces a larger fraction of gaseous products. Exothermic effects within passive pyrolysis stage can be attributed to decomposition patterns of hemicelluloses and lignin that are related to their high charring nature, and which is in contrast with decomposition of cellulose taking place by full volatilization [37]. Bio-char consists of residue of lignin [38, 39], where lignin continues to further exothermic polymerization stage of bio-char. At higher temperatures, polymerization of biomass bio-char continuously occurred and polymerization reaction is dependent on unique properties of bio-char of selected biomass sample.

3.3 Analysis of Gases Produced During Pyrolysis

The thermogravimetric analysis—mass spectrometry (TGA-MS) profiles of the evolved gases during pyrolysis, H_2, CH_4, C_2H_6, C_4H_{10}, C_5H_{12} and profiles of the changes in the ion current as a function of temperature (the mass spectra) for pyrolytic water (H_2O), CO_2 (C_3H_8) and CO for various biomasses (HS – hazelnut shell, SD-sawdust, and SDCT – sawdust chemically treated) at the different heating rates of $\beta = 5$, 10 and 20 °C min^{-1} are shown in Figs. 4, 5 and 6(a)–(f). The present study was focused on main volatile products of pyrolysis on the basis of changes in ion currents across the temperature range of 40–600 °C, and on their relevancy, assigned to the ion/mass intensities (m/z) with corresponded screening analysis performed in the selected-ion monitoring (SIM) mode (the ions characteristic of each molecules, such as: 2, 16, 18, 28, 44, … for H_2, CH_4, H_2O, CO, CO_2, … were monitored) and which is in accordance with database of National Institute of Standards and Technology (NIST).

At all observed heating rates, the primary devolatilization stage was identified with major mass loss and the release of organic compounds which leads to formation bio-char, where this stage releases H_2, CH_4, H_2O, CO_2 (also C_3H_8), CO, including the light hydrocarbons. The maximums in DTG curves (Fig. 2) correspond to the maximum gas releases, as shown in Figs. 4, 5 and 6. The largest deviation shows the CO evolution profile for hazelnut shell (HS) pyrolysis in respect to DTG maxima features, and this deviation is particularly pronounced, when changing the heating rate from lowest to highest rate of the heating. This is consequence of non-profiled DTG peak, but a clear appearance of "shoulder", which in many cases depends on hemicelluloses concentration (if the concentration is low).

For the sawdust and MDF sawdust, the gas evolution profiles (also except for CO) are relatively symmetrical ones and follow the DTG profiles. Having in mind these results and comparing the CO gas profiles (including CO_2 profile), it can be concluded that contribution of lignin plays a significant role and its contribution largely depends on the rate of heating. Since the lignin is decomposed continuously in a wide temperature range, with un-defined rate of the mass loss, such behavior is manifested through a series of "unsteady" profiles of CO evolution. It should be noted that for the cases observed (in particular, for sawdust pyrolysis), the CH_4 and H_2 peaks are detected in passive pyrolysis zone for $T > 450$ °C. Such secondary pyrolysis is result of decomposition of heavy molecules in bio-char. The H_2 evolved is, in fact, only a part of hydrogen present due to the higher volatile matter content in the samples. With regard to H_2, the charring and restructuring processes, which include dehydrogenation reac-

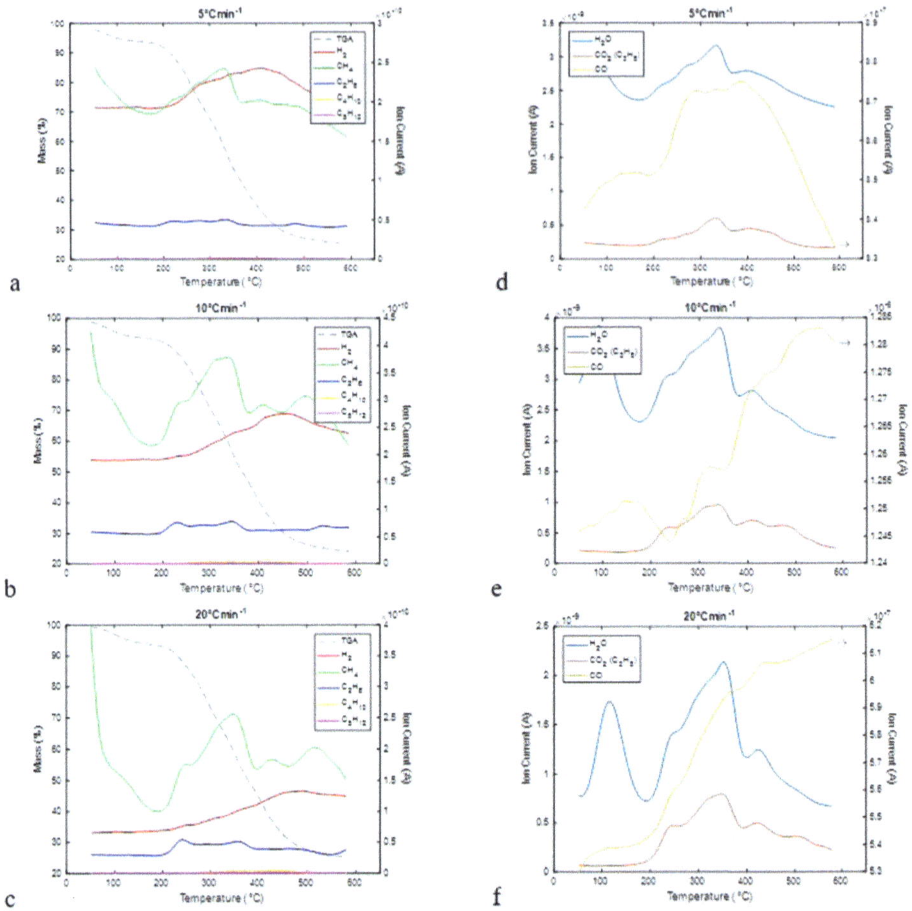

Fig. 4. Simultaneous *TGA-MS profiles of evolved gases during pyrolysis of hazelnut shell*

tions, are responsible for its production, starting at temperature above 450 °C, depending on the sample. The CH_4, as well as other C_2 and C_3 hydrocarbons, present local maxima at temperatures different than ones of the maximum of DTG curve.

For HS, two local maxima at 10 and 20 °C min^{-1} in temperature ranges 400–450 °C /450–550 °C correspond to methane evolution peaks. The lower temperature peak can be attributed to $-OCH_3$ functional group cracking, while the higher temperature peak is due to charring processes [40]. For SD and SDCT, it can be identified one local maximum for methane evolution peak in temperature range of 380–475 °C (except for SD at 20 °C min^{-1}, where two local maxima appear). A relatively significant amount of CO and H_2O are detected because of large number of hydroxyl groups and oxygen atoms present in tested biomass samples.

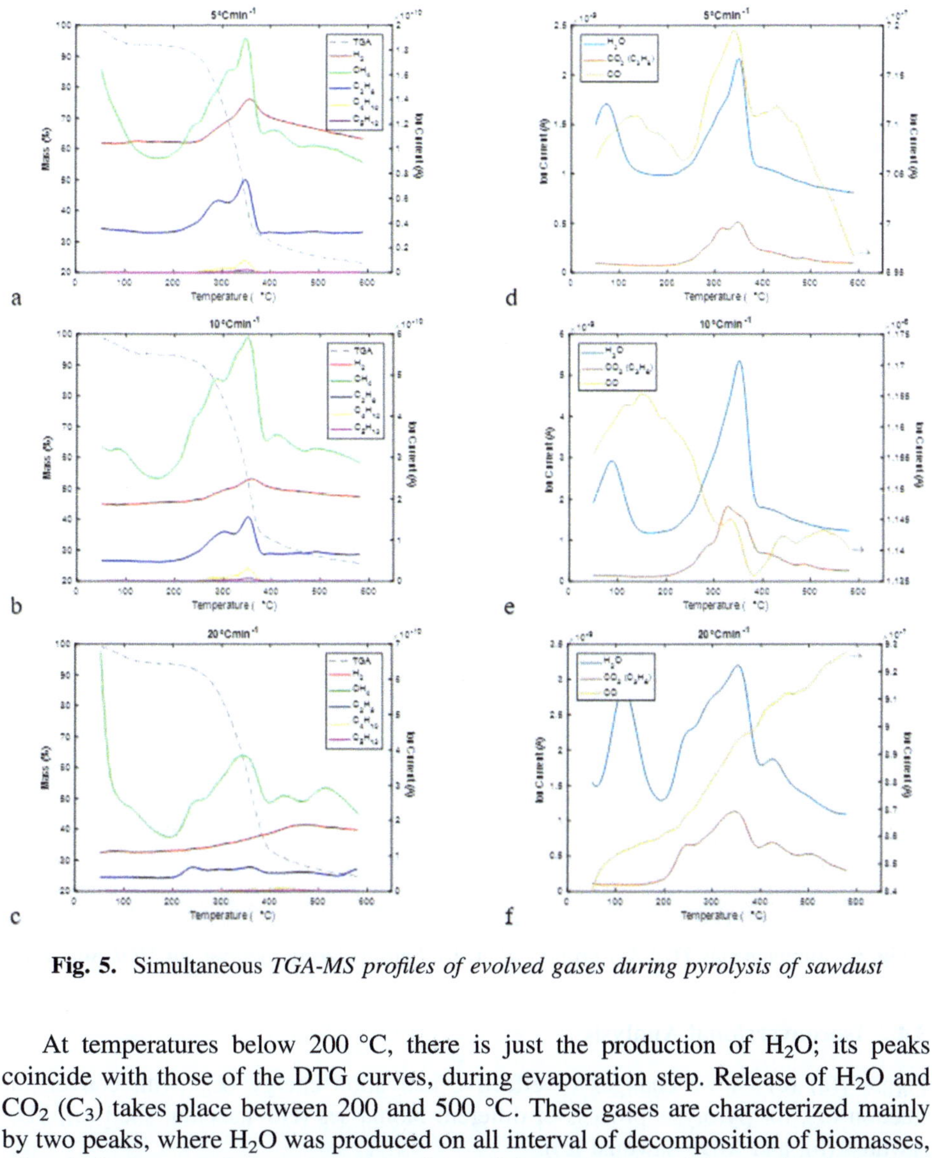

Fig. 5. Simultaneous *TGA-MS profiles of evolved gases during pyrolysis of sawdust*

At temperatures below 200 °C, there is just the production of H_2O; its peaks coincide with those of the DTG curves, during evaporation step. Release of H_2O and CO_2 (C_3) takes place between 200 and 500 °C. These gases are characterized mainly by two peaks, where H_2O was produced on all interval of decomposition of biomasses, what indicates on their provenance from different origins. The peak obtained at approximately at 105 °C is due to liberation of absorbed water, while peaks beyond 200 °C correspond to the water formed during the decomposition of hemicelluloses, cellulose and lignin [41]. The productions of C_2H_6 results from decomposition of cellulose while the release of CH_4 at low temperatures, comes from decompositions of cellulose and hemicelluloses [42]. At high temperatures, CH_4 is mainly produced by cracking of the methoxyl groups [43].

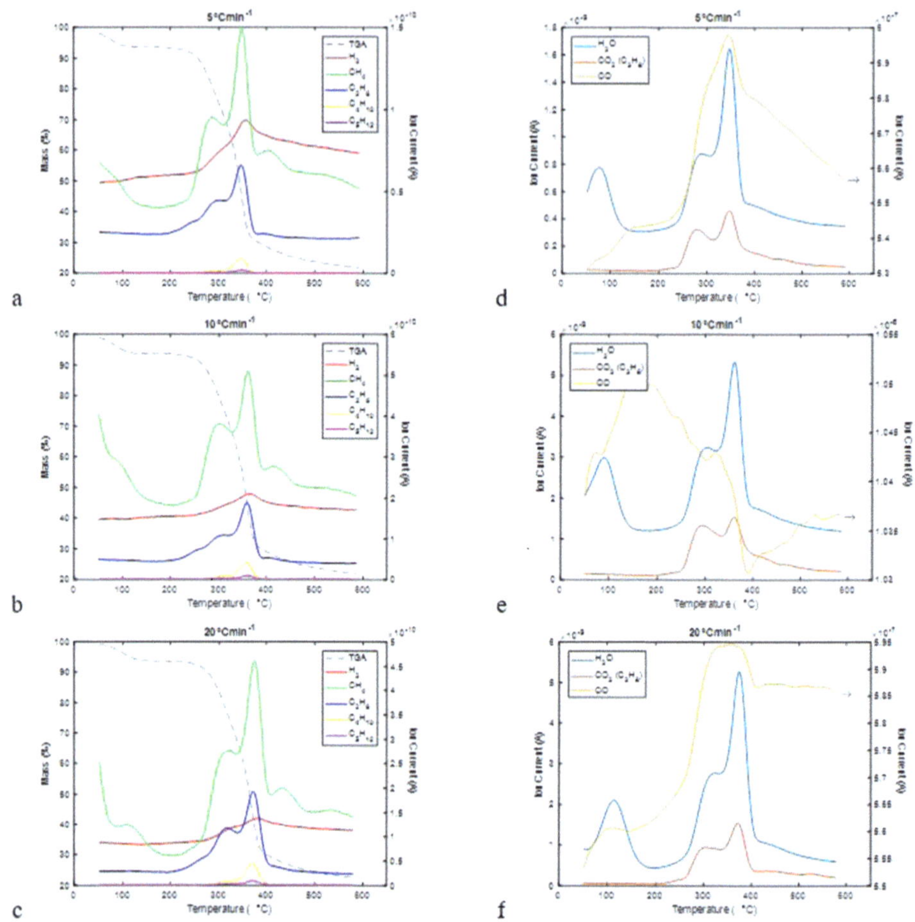

Fig. 6. Simultaneous *TGA-MS profiles of evolved gases during pyrolysis of MDF sawdust*

3.4 Isoconversional Analysis

Figure 7(a)–(c) shows changes of effective activation energy (E_a) with conversion fraction (α), for pyrolysis process of different biomasses (HS–hazelnut shell (a), SD–sawdust (b), and MDF sawdust (c)).

In all considered cases, the effective activation energy change curves clearly, so this illustrates that pyrolysis processes of biomasses are too complex to be characterized by a single-step reaction. The detailed evaluation of E_a vs. α curves, indicates that only for HS pyrolysis, the E_a values are unique and almost constant at certain conversion intervals, as $E_{a,avg}$ = 227.7 kJ mol^{-1} (for α = 0.40–0.65, ΔT = 305–345 °C) (Fig. 7 (a)). For sawdust and MDF sawdust pyrolysis, such conversion intervals, in which E_a is almost constant, cannot be separated, because it exists a very large magnitude variation in E_a's during entire pyrolysis (Fig. 7(b) and (c)).

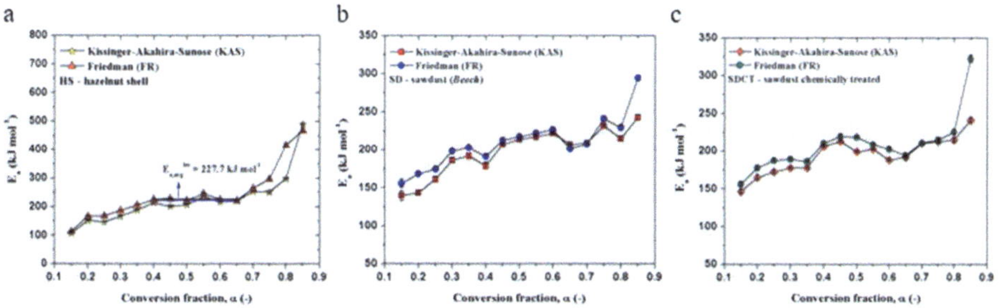

Fig. 7. Effective activation energy (via KAS and FR methods) versus conversion fraction for (a) HS–hazelnut shell, (b) SD–sawdust and (c) MDF-sawdust pyrolysis processes

Considering estimated average values of effective activation energy ($E_{a,avg}$) for CB, WS and HS, we can conclude that these values may be attributed to the cellulose decomposition, where decomposition of cellulose generally took place in detected conversion/temperature intervals. Namely, through literature review, the average value of E_a = 198 kJ mol^{-1} for pure cellulose, and the value of E_a = 228 kJ mol^{-1} for microcrystalline cellulose were found [44, 45]. It can be observed that E_a values climbed rapidly at very end of the process ($\alpha > 0.75$). The E_a is an obstacle that must be overcome before a chemical reaction is generated, and higher value of E_a implies more difficult for a reaction to occur. It determines the reactivity and sensitivity of a reaction rate [46]. It is likely that minerals present in biomasses act as a barrier for the diffusion of the heat and the release of decomposed volatiles in charring process. The higher values of E_a identified in all cases at the end of pyrolysis processes can also be attributed to an increase in thermal stability as a result of the increasing of aromatic character of the lignin-derived bio-char when higher temperatures are reached, turning lignin into a highly cross-linked carbonaceous material.

However, it was announced that multiple causes (e.g., experimental and calculation system errors, thermal lag, temperature gradient, compensation effect, etc.) are at origin of these variations and led to major confusion in interpretation of biomass pyrolysis kinetics. However, the consistent pattern for biomass decomposition is often observed and generally described by global reactions, including decomposition of three pseudo-components, with typical effective activation energies of 57–187 kJ mol^{-1} for hemicelluloses decomposition [47], 195–236 kJ mol^{-1} for cellulose decomposition [48], and 35–267 kJ mol^{-1} for lignin decomposition [14]. Present variations of E_a with changing of α value arise from different chemical structures among cellulose, hemicelluloses and lignin in tested samples and which may affect their effective activation energies. Furthermore, amorphous nature of hemicelluloses causes it to be less stable than other two pseudo-components. Due to its strong inter-molecular hydrogen bonding, cellulose required higher effective activation energy than hemicelluloses.

4 Conclusions

This work presented an experimental and kinetic study of pyrolysis process of various biomasses (hazelnut shell (HS), sawdust (SD), and sawdust chemically treated (SDCT)) using simultaneous thermal analysis (STA) – mass spectrometry (MS) techniques, in an inert atmosphere. According to performed analysis following conclusions can be drawn.

TGA and DTG results showed that pyrolysis of the samples were represented by three successive steps: the first step corresponded to the evaporation of moisture, while the second and third steps can be attributed to active and passive pyrolysis, respectively. Active pyrolysis stage is characterized by the largest mass loss where various vapors and gases are released as a result of decomposition processes of hemicelluloses and cellulose. The third pseudo-component of biomass (lignin) is decomposed in both stages: active and passive pyrolysis stages, without characteristic thermo-analytical (TA) peaks. It was found that certain variations in the heat effects can be significantly stimulated by concentration of three pseudo-components, which also affects development of theirs decomposition paths. It was established that differences in molecular structures and chemical nature of three pseudo-components in biomasses should be accounted for dissimilarities observed in DTG curves.

MS results showed that H_2, CH_4, H_2O, CO_2 (C_3H_8), CO, and C_2H_6 were main gaseous products released during pyrolysis. It was found that majority of gases is generated in temperature range 200–500 °C, where differences occur in respect to changing in heating rate and in MS peak intensity. It was found that H_2O, CO and CO_2 evolutions for all biomasses arise from lignin source in biomass, followed by cellulose, and hemicelluloses. Considering biomass pseudo-components pyrolysis, it was found that CH_4 was mainly released from lignin pyrolysis, because it contains more methoxyl-O-CH_3 chemical groups than hemicelluloses and celluloses. It was found that at temperatures higher than 450 °C, the H_2 formation can be attributed to secondary pyrolytic reactions. At lower temperatures, the H_2 emission was mainly generated from thermal decomposition of cellulose and hemicelluloses.

The higher effective activation energies were identified in all cases. This behavior can be attributed to an increase in thermal stability as a result of increasing of aromatic character of lignin-derived bio-char when higher temperatures are reached.

Acknowledgments. Authors would like to acknowledge financial support of Ministry of Education, Science and Technological Development of the Republic of Serbia under the Projects III42010 and 172015.

References

1. Ohlström, M., Mäkinen, T., Laurikko, J., Pipatti, R.: New concepts for biofuels in transportation: biomass-based methanol production and reduced emissions in advanced vehicles. VTT Research Note 3-94, VTT (Technical Research Centre of Finland) Energy, Espoo (2001)
2. Saxena, R.C., Adhikari, D.K., Goyal, H.B.: Biomass-based energy fuel through biochemical routes: a review. Renew. Sustain. Energy Rev. **13**, 167–178 (2009). https://doi.org/10.1016/j.rser.2007.07.011

3. Maniatis, K.: Progress in biomass gasification: an overview. Prog. Thermochem. Biomass Convers. (2001). https://doi.org/10.1002/9780470694954.ch1
4. Faaij, A.: Modern biomass conversion technologies. Mitig. Adapt. Strateg. Glob. Chang. **11**, 343–375 (2006). https://doi.org/10.1007/s11027-005-9004-7
5. Anwar, Z., Gulfraz, M., Irshad, M.: Agro-industrial lignocellulosic biomass a key to unlock the future bio-energy: a brief review. J. Radiat. Res. Appl. Sci. **7**, 163–173 (2014). https://doi.org/10.1016/j.jrras.2014.02.003
6. Garcia-Perez, M., Lewis, T., Kruger, C.E.: Methods for producing biochar and advanced biofuels in Washington State, Part 1: Literature review of pyrolysis reactors. First project report, Department of Biological Systems Engineering and the Center for Sustaining Agriculture and Natural Resources Washington State University, Pullman (2010)
7. Czajczyńska, D., Anguilano, L., Ghazal, H., Krzyżyńska, R., Reynolds, A.J., Spencer, N., Jouhara, H.: Potential of pyrolysis processes in the waste management sector. Therm. Sci. Eng. Prog. **3**, 171–197 (2017). https://doi.org/10.1016/j.tsep.2017.06.003
8. Jayaraman, K., Gökalp, I.: Pyrolysis, combustion and gasification characteristics of miscanthus and sewage sludge. Energy Convers. Manag. **89**, 83–91 (2015). https://doi.org/10.1016/j.enconman.2014.09.058
9. Alipour Moghadam, R., Yusup, S., Azlina, W., Nehzati, S., Tavasoli, A.: Investigation on syngas production via biomass conversion through the integration of pyrolysis and air-steam gasification processes. Energy Convers. Manag. **87**, 670–675 (2014). https://doi.org/10.1016/j.enconman.2014.07.065
10. Dhaundiyal, A., Singh, S.B., Hanon, M.M., Rawat, R.: Determination of kinetic parameters for the thermal decomposition of Parthenium hysterophorus. Environ. Clim. Technol. **22**(1), 5–21 (2018)
11. Wang, L., Lei, H., Liu, J., Bu, Q.: Thermal decomposition behavior and kinetics for pyrolysis and catalytic pyrolysis of Douglas fir. RSC Adv. **8**, 2196–2202 (2018). https://doi.org/10.1039/C7RA12187C
12. Janković, B., Marinović-Cincović, M., Janković, M.: TG-DTA-FTIR analysis and isoconversional reaction profiles for thermal and thermo-oxidative degradation processes in black chokeberry (*Aronia melanocarpa*). Chem. Pap. **70**(8), 1094–1105 (2016). https://doi.org/10.1515/chempap-2016-0046
13. Janković, B.: The comparative kinetic analysis of Acetocell and Lignoboost® lignin pyrolysis: the estimation of the distributed reactivity models. Bioresour. Technol. **102**(20), 9763–9771 (2011). https://doi.org/10.1016/j.biortech.2011.07.080
14. Várhegyi, G., Antal, M.J., Jakab, E., Szabó, P.: Kinetic modeling of biomass pyrolysis. J. Anal. Appl. Pyrolysis **42**, 73–87 (1997). https://doi.org/10.1016/s0165-2370(96)00971-0
15. EN ISO 14780:2017: Solid biofuels – sample preparation. European Committee for Standardization (CEN), Brussels (2017)
16. EN ISO 18134:2015: Solid biofuels – determination of moisture content – oven dry method – Part 1: Total moisture – reference method. European Committee for Standardization (CEN), Brussels (2015)
17. EN ISO 16948:2015: Solid biofuels – determination of total content of carbon, hydrogen and nitrogen. European Committee for Standardization (CEN), Brussels (2015)
18. EN ISO 17225-1:2014: Solid biofuels – fuel specifications and classes – Part 1: General requirements. European Committee for Standardization (CEN), Brussels (2015)
19. EN ISO 18125:2017: Solid biofuels – determination of calorific value. European Committee for Standardization (CEN), Brussels (2015)
20. Grønli, M.G., Várhegyi, G., Di Blasi, C.: Thermogravimetric analysis and devolatilization kinetics of wood. Ind. Eng. Chem. Res. **41**(17), 4201–4208 (2002)

21. Hu, M., Wang, X., Chen, J., Yang, P., Liu, C., Xiao, B., Guo, D.: Kinetic study and syngas production from pyrolysis of forestry waste. Energy Convers. Manag. **135**, 453–462 (2017). https://doi.org/10.1016/j.enconman.2016.12.086
22. Martín-Lara, M.A., Blázquez, G., Ronda, A., Calero, M.: Kinetic study of the pyrolysis of pine cone shell through non-isothermal thermogravimetry: effect of heavy metals incorporated by biosorption. Renew. Energy **96**, 613–624 (2016). https://doi.org/10.1016/j.renene.2016.05.026
23. Kissinger, H.E.: Reaction kinetics in differential thermal analysis. Anal. Chem. **29**, 1702–1706 (1957). https://doi.org/10.1021/ac60131a045
24. Akahira, T., Sunose, T.: Method of determining activation deterioration constant of electrical insulating materials. Res. Rep. Chiba Inst. Technol. (Sci. Technol.) **16**, 22–31 (1971)
25. Friedman, H.L.: Kinetics of thermal degradation of char-forming plastics from thermogravimetry. Application to a phenolic plastic. J. Polym. Sci. Part C Polym. Symp. **6**, 183–195 (2007). https://doi.org/10.1002/polc.5070060121
26. Venkatesh, M., Ravi, P., Tewari, S.P.: Isoconversional kinetic analysis of decomposition of nitroimidazoles: Friedman method vs Flynn–Wall–Ozawa Method. J. Phys. Chem. A **117** (40), 10162–10169 (2013). https://doi.org/10.1021/jp407526r
27. Demirbas, A.: Hydrogen from mosses and algae via pyrolysis and steam gasification. Energy Sources Part A Recover. Util. Environ. Eff. **32**, 172–179 (2009). https://doi.org/10.1080/15567030802464388
28. Mohan, D., Pittman, C.U., Steele, P.H.: Pyrolysis of wood/biomass for bio-oil: a critical review. Energy Fuels **20**, 848–889 (2006). https://doi.org/10.1021/ef0502397
29. Hellier, P., Talibi, M., Eveleigh, A., Ladommatos, N.: An overview of the effects of fuel molecular structure on the combustion and emissions characteristics of compression ignition engines. Proc. Inst. Mech. Eng. Part D J. Automob. Eng. **232**(1), 90–105 (2018). https://doi.org/10.1177/0954407016687453
30. Song, H., Quinton, K.S., Peng, Z., Zhao, H., Ladommatos, N.: Effects of oxygen content of fuels on combustion and emissions of diesel engines. Energies **9**, 28–39 (2016). https://doi.org/10.3390/en9010028
31. McKendry, P.: Energy production from biomass (part 1): overview of biomass. Bioresour. Technol. **83**, 37–46 (2002). https://doi.org/10.1016/s0960-8524(01)00118-3
32. Ronsse, F., van Hecke, S., Dickinson, D., Prins, W.: Production and characterization of slow pyrolysis biochar: influence of feedstock type and pyrolysis conditions. GCB Bioenergy. **5**, 104–115 (2012). https://doi.org/10.1111/gcbb.12018
33. Brownsort, P.A.: Biomass pyrolysis processes: review of scope, control and variability. Edinburgh Biochar Research Centre working paper 5, pp. 1–39, The University of Edinburgh, Newcastle University, Rothamsted Research, December 2009. www.biochar.org.uk
34. Conesa, J.A., Caballero, J., Marcilla, A., Font, R.: Analysis of different kinetic models in the dynamic pyrolysis of cellulose. Thermochim. Acta **254**, 175–192 (1995). https://doi.org/10.1016/0040-6031(94)02102-t
35. Caballero, J.A., Conesa, J.A., Font, R., Marcilla, A.: Pyrolysis kinetics of almond shells and olive stones considering their organic fractions. J. Anal. Appl. Pyrolysis **42**, 159–175 (1997). https://doi.org/10.1016/s0165-2370(97)00015-6
36. Janković, B.: Kinetic and reactivity distribution behaviors during curing process of carbon/epoxy composite with thermoplastic interface coatings (T800/3900–2 prepreg) under the nonisothermal conditions. Polym. Compos. **39**(1), 201–220 (2018). https://doi.org/10.1002/pc.23920
37. Yang, H., Yan, R., Chen, H., Lee, D.H., Zheng, C.: Characteristics of hemicellulose, cellulose and lignin pyrolysis. Fuel **86**, 1781–1788 (2007). https://doi.org/10.1016/j.fuel.2006.12.013

38. Sonobe, T., Pipatmanomai, S., Worasuwannarak, N.: Pyrolysis characteristics of Thai-agricultural residues of rice straw, rice husk, and corncob by TG-MS technique and kinetic analysis. In: Proceedings of 2nd Joint International Conference on Sustainable Energy and Environment (SEE 2006), 21–23 Nov 2006, Bangkok, Thailand, C-044, pp. 1–6 (2006)
39. Domingues, R.R., Trugilho, P.F., Silva, C.A., de Melo, I.C.N.A., Melo, L.C.A., Magriotis, Z.M., Sánchez-Monedero, M.A.: Properties of biochar derived from wood and high-nutrient biomasses with the aim of agronomic and environmental benefits. PLoS ONE **12**(5), e0176884 (2017). https://doi.org/10.1371/journal.pone.0176884
40. Gómez, C.J., Mészáros, E., Jakab, E., Velo, E., Puigjaner, L.: Thermogravimetry/mass spectrometry study of woody residues and an herbaceous biomass crop using PCA techniques. J. Anal. Appl. Pyrolysis **80**, 416–426 (2007). https://doi.org/10.1016/j.jaap.2007.05.003
41. Janković, B.: Estimation of the distribution of reactivity for powdered cellulose pyrolysis in isothermal experimental conditions using the Bayesian inference. Cellulose **22**(4), 2283–2303 (2015). https://doi.org/10.1007/s10570-015-0653-8
42. Özveren, U., Özdoğan, Z.S.: Investigation of the slow pyrolysis kinetics of olive oil pomace using thermo-gravimetric analysis coupled with mass spectrometry. Biomass Bioenerg. **58**, 168–179 (2013). https://doi.org/10.1016/j.biombioe.2013.08.011
43. Jakab, E., Faix, O., Till, F.: Thermal decomposition of milled wood lignins studied by thermogravimetry/mass spectrometry. J. Anal. Appl. Pyrolysis **40–41**, 171–186 (1997). https://doi.org/10.1016/s0165-2370(97)00046-6
44. Lin, Y.-C., Cho, J., Tompsett, G.A., Westmoreland, P.R., Huber, G.W.: Kinetics and mechanism of cellulose pyrolysis. J. Phys. Chem. C **113**, 20097–20107 (2009). https://doi.org/10.1021/jp906702p
45. Antal, M.J., Várhegyi, G., Jakab, E.: Cellulose pyrolysis kinetics: revisited. Ind. Eng. Chem. Res. **37**, 1267–1275 (1998). https://doi.org/10.1021/ie970144v
46. Janković, B.: The pyrolysis process of wood biomass samples under isothermal experimental conditions—energy density considerations: application of the distributed apparent activation energy model with a mixture of distribution functions. Cellulose **21**(4), 2285–2314 (2014). https://doi.org/10.1007/s10570-014-0263-x
47. Orfão, J.J.M., Antunes, F.J.A., Figueiredo, J.L.: Pyrolysis kinetics of lignocellulosic materials—three independent reactions model. Fuel **78**, 349–358 (1999). https://doi.org/10.1016/s0016-2361(98)00156-2
48. Grønli, M.G., Várhegyi, G., Di Blasi, C.: Thermogravimetric analysis and devolatilization kinetics of wood. Ind. Eng. Chem. Res. **41**, 4201–4208 (2002). https://doi.org/10.1021/ie0201157

Thermal Energy Storage of Composite Materials Based on Clay, Stearic Acid, Paraffin and Glauber's Salt as Phase Change Materials

Milena Stojiljkovic[1(✉)], Stanisa Stojiljkovic[1], Bratislav Todorovic[1], Mirjana Reljic[2], Sasa Savić[1], and Sanja Petrovic[1]

[1] Faculty of Technology, University of Nis, Bulevar Oslobodjenja 124, 16000 Leskovac, Serbia
milena0919@live.com
[2] CIS Institute, Vojislava Ilica 88, Belgrade, Serbia

Abstract. Thermal protection and insulation are important problems in many fields such as industry, agriculture and medicine. New composite materials with good thermal storage capacities have become important in the last few decades. The role of these materials is reflected in their ability to store energy and allow it to be reused in some other thermal systems. The aim of this study was to create a new material based on the basically activated bentonite clay. First, the clay was basically activated, resulting in a thick gel. Afterwards, stearic acid, Glauber's salt and active carbon were added, and a heterogeneous gel was obtained as a finished final product. In order to obtain the best heterogeneous gel with satisfactory storage properties, the amount of stearic acid and Glauber's salt was varied. The characterization of the resulting heterogeneous gel was performed by measuring the cooling rate of the gel samples. Compared with stearic acid, Glauber's proved to be more effective. Heterogeneous gel cooling tests have shown that there was a certain proportional dependence between the concentration of stearic acid and the Glauber salt. However, it has been noticed the reduction in the cooling rate. Namely, the increase in stearic acid and Glauber's salt concentration lead to slowing down the cooling rate of the gel. Adding active carbon to the heterogeneous gel also reduced the cooling rate, which indicated that the presence of active carbon in the heterogeneous gel should not be excluded in the future. The advantage of this system is the improvement of the gel thermal characteristics by the presence of water and clay. The gel was reversibly cooled and heated up to 100 °C without changing the homogeneous structure. This system can be used as a heat recovery pad, due to its flexible body pillow. It can be very quickly warmed up in a microwave oven if it is packaged in polyethylene packaging.

Keywords: Thermal storage · Bentonite clay · Glauber's salt
Paraffin · Stearic acid · Activated carbon

N. Mitrovic et al. (Eds.): CNNTech 2018, LNNS 54, pp. 34–43, 2019.
https://doi.org/10.1007/978-3-319-99620-2_3

1 Introduction

Thermal energy storages (TESs) in the last few decades become very popular, especially due to their important role in the processes of heat recovery processes as well as in improving the performance of the thermal systems [1–3]. TES contain mediums able to store the thermal energy (charging mode) and release the stored thermal energy (discharging mode) when it is necessary to compensate deficiency in the main thermal source [4–6]. The most common storage medium is a fluid or phase change material (PCM). National Aeronautics and Space Administration (NASA) has reported seven material categories with phase change properties: paraffins, non-paraffin organics, salt hydrates, metallic, fused salt eutectic mixtures, miscellaneous, and solid-solid [7, 8]. Each of these materials possesses some advantages and limitations, so the selection has to be done based on the application requirements [9, 10]. Generally, all these materials have a low thermal conductivity which induces slow charging and discharging rate [11]. In order to eliminate the existing lacks, scientists began to use nano-additives to enable continuous and linear energy exchange. Nanomaterials are used as additives to change the basic materials characteristics [12–14]. Once added to the fluid, produce mixture become nanofluids. Nanomaterials additions to the phase change materials provide nano-composites. Typical paraffin waxes are saturated hydrocarbon mixtures and usually consist of a mixture predominantly of alkanes of the real molecular chains such as CH_3-(CH_2)n-CH_3. Crystallization (CH_2)n chains can liberate a large amount of latent heat. Fatty acids are characterized by the formula $CH_3(CH_2)_{2n}COOH$. Paraffin waxes and fatty acids are organic supplements and are distinguished from inorganic PCM materials. As they are mostly chemically inert, stable, easy to recycle, these compounds show phase separation and corrosive behavior (with the exception of fatty acids). However, some disadvantages are low thermal conductivity, volumetric storage (less than 103 kg/m^3), and flammability, unlike inorganic PCM materials [15, 16].

Results obtained by scanning electron microscopy (SEM) showed that the bentonite clay represent a good structural barrier for the organic PCMs homogenously dispersed onto its surface and interlayers [17, 18]. The chemical investigations made by using Fourier transform infrared (FT-IR) spectroscopy revealed that the attractions between the components of the composites were physical in nature and thus the PCM were hold by capillary forces [19, 20]. The advantage of this system is that the storage temperature may be higher than 100 °C, while pressure still remains very close to atmospheric pressure. Activations of clay by alkalis allows homogenization of the materials capable of melting heat change, whether an organic or inorganic nature [21, 22].

In the process of composites production, achieving the stable thermal cycles without loss of medium mass is the main goal. By combining the relationship of the Glauber's salt and stearic acid into the selected composite, it is possible to influence the cooling kinetics. By increasing the content of the Glauber salt, it is possible to cool at temperatures close to its melting point. These are temperatures between 40 °C and 50 °C [23, 24].

In addition to clay, activated carbon in composite structure is also popular. It has porous structure with many inner surfaces, which allows it to be easily saturated with the melted PCM [6, 13, 25]. The activated carbon is suitable to form carbonaceous layers, which creates a physical protective barrier on the composites surface. The protective barrier could confine the transfer of flammable molecules to the gas phase. This results in the improvement of composites thermal stability [12, 14, 26]. The high thermal conductivity of the graphite ligaments in the foam allows rapid transfer of heat throughout the PCM volume. The PCM is able to absorb a significant amount of heat without a great increase in temperature during the phase change [27–29]. The aim of this study is to make a composite by using the bentonite clay, Glauber's salt, paraffin wax, stearic acid and activated carbon in order to determine the composite mixture with the largest heat storage capacity of 100 °C.

In this study we have integrated modified bentonite clay which in combination with water forms a gel miscible with a PCM. The resulting composite is nanofluid gel, with very pronounced thixotropy [30–32]. PCMs usually have a rather small values of thermal conductivity, in range of 0.1 to 0.3 W/(mK), but in combinations with other compounds their thermal conductivity could be significantly improved [11, 27].

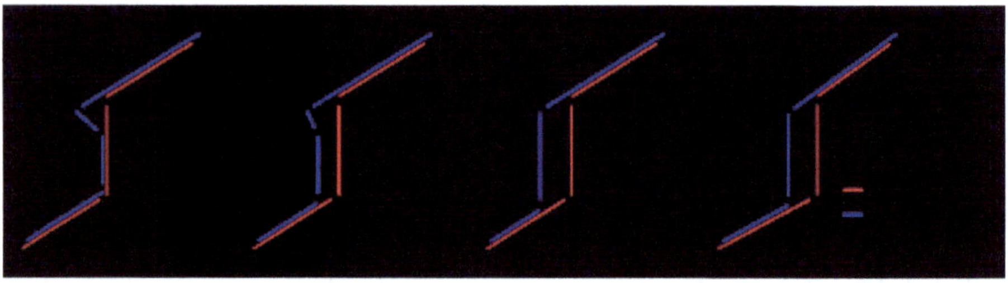

Fig. 1. Real hysteresis as a material property caused by subcooling when: (a) the temperature rises again to the solidifying temperature of the PCM; (b) the temperature does not rise again to the solidifying temperature of the PCM. (c) Real hysteresis caused by slow heat release or a real difference between the phase-change temperatures. (d) Apparent hysteresis caused by non-isothermal conditions in the measurements [33].

PCMs have a high density of storage of heat in a small temperature range. In this case, this is called apparent hysteresis. Figure 1 shows the different causes that can lead to hysteresis. Due to hysteresis, there are usually different data from charging and discharging experiments. Another aspect to be taken into account when it comes to heat transfer with solid-liquids phase change. In melted PCM is the effect of natural convection, because it is one of the main factors that influence the processes of phase transition. PCMs need to show a repeatable phase change in order to maintain the storage capacity of the heat needed for the designed application. It should be noted PCM showing phase separation will show a decrease in the enthalpy of melting after repetition cycling. For PCM-based elements, stability in cycling may also refer to the ability to avoid fluid leak after repeated melting and solidification cycles [33].

2 Experimental

2.1 Materials

Bentonites clay was obtained from locations of Prisjan in the Pirot district (Serbia). Its chemical composition after drying at 110 °C is as follows: 51.82% SiO_2, 0.34% TiO_2, 26.86% Al_2O_3, 2.30% Fe_2O_3, 0.10% MnO, 1.27% MgO, 1.44% CaO, 0.75% Na_2O and 2.07% K_2O. Borax, Glauber's salt, paraffin and stearic acid were obtained from ARIHEM (Nis, Serbia), while activated carbon was made by Miloje Zakic d.o.o. (Krusevac, Serbia). Temperature measurement was done using Digital thermometer DT-1 (Dalmacija, Dugi rat, Croatia).

2.2 Heterogeneous Gel Based on Alkali Bentonite Clay Activation Making Procedure

In the first phase, 100 ml of water was heated to the temperature of 90 °C and 60 g of bentonite clay was added. After intensive stirring, obtained liquid was concentrated by addition of 5 g NaOH. In the second phase, in easy flowable gel that was previously obtained by stirring activated carbon, Glauber's salt, paraffin and stearic acid were added in appropriate amounts. In the last phase after the intensive stirring, easily mobile and homogeneous thixotropic gel was obtained, in which organic phase with the inorganic was emulsified.

2.3 Characterization of the Heterogeneous Gel

Heterogeneous gel in test tubes was heated in a temperature-regulated water bath to the temperature of 95 °C, and after completely melting the samples, the test tubes were removed from the hot water bath and left in an isolating vessel made of polystyrene (isolation diameter was 10 cm) which is shown in Fig. 2. All off experiments were done at ambient temperature for samples cooling to the room temperature. A temperature loss is measured every 60 min for a period of 9 h. In order to analyze the cooling

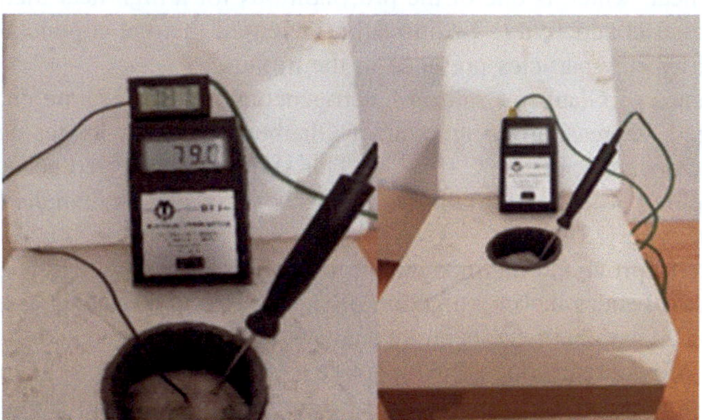

Fig. 2. Laboratory model for heat storage determination

behavior and the cooling degree of each sample, graphs of temperature changes during the time were plotted.

3 Results and Discussion

Results indicate a proportional decreasing of cooling speed with increasing of mass of Glauber's salt in heterogeneous gel. At the relative ratio of Glauber's salt and stearic acid of 1:2, the temperature storage effect is lower for about 5 to 10% compared to the 2:1 ratio in favor of Glauber's salt (Figs. 1 and 2). According to results, Glauber's salt is important for cooling speed decrease. Based on the obtained results it can be concluded that Glauber's salt amount increasing may affect more favorable conditions for the heat storage.

Results of the stearic acid influence on temperature changes of heterogeneous gel are presented in a Fig. 3. Obtained results indicate that stearic acid has effect on cooling of heterogeneous gel, but that effect is not significant. Trend of temperature decreasing at stearic acid concentrations of 5, 10 and 15 g is almost identical, while at the highest stearic acid concentration (20 g) cooling trend of heterogeneous gel is lower compared to the previous investigated concentrations [34].

Figure 1(a) and (b) shows the phase transitions with subcooling, which is reduced by the introduction of expanded clay. The gel based on expanded clay with water is high pH value. It provides a mild phase transition without subcooling [33].

Generally, thermal conductivity increase has positive influence on thermal storage properties of composite material. Namely, composite material will be effective as thermal storage material if during one cycle is fully charged (all PCM should melt) and then fully uncharged (all PCM should solidify), while the process of charging and discharging depends on the thermal conductivity of the composite. Clay activation by alkali in mixture with water creates a gel that has strong thixotropic properties. With minimal mixing, the gel becomes a fluid and easily maneuverable, that after a while becomes solid consistency. In this combination the presence of water induces primarily high specific heat [15]. It is known that water has a high coefficient of conduction and high specific heat, which is one of the preconditions for a high heat mixture capacity and high thermal conductivity. Additionally, increased thermal conductivity could be also achieved by clay particles presence in the mixture.

The influence of Glauber's salt on heterogeneous gel temperature changes is presented in Fig. 4. Increasing the amount of Glauber's salt is a longer period of heat transfer at a temperature of 50 °C (Fig. 4). This should be expected because the latent heat of the melting (vaporization) of the Galuber's salt is greater than the latent heat of stearic acid melting [23].

In order to improve the thermodynamic properties in nanofluid gel, activated carbon was added. Besides the investigation of stearic acid and Glauber's salt influence, effect of activated carbon on temperature changes was also investigated. It can be noted that a sample of heterogeneous gel with active carbon has proven to be more effective because of a slower cooling system for heat storage. For nine hours the heterogeneous gel temperature dropped from 95 to 48.2 °C, which is for about 10 °C higher compared to the heterogeneous gel in which activated carbon was not present (Fig. 5).

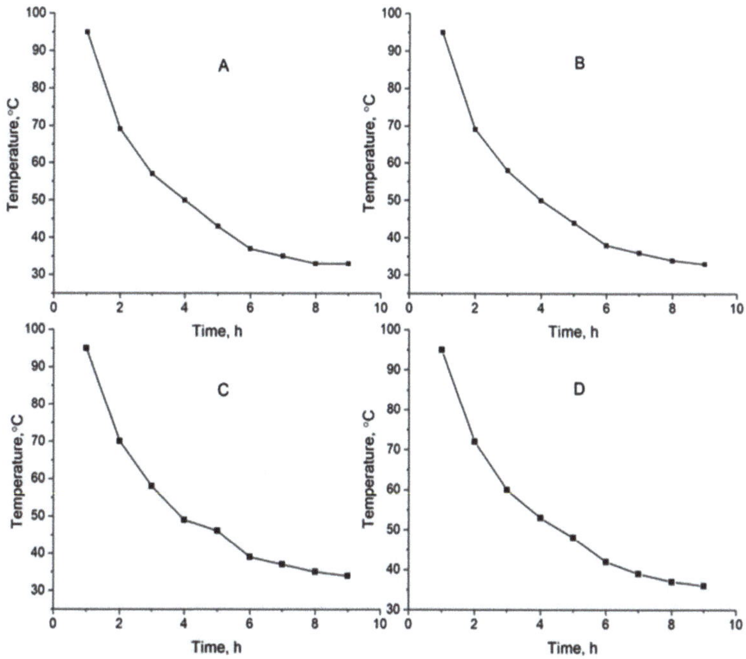

Fig. 3. Influence of stearic acid mass on heterogeneous gel temperature changes. Heterogeneous gel composition: 100 g of water, 60 g of clay, 5 g of NaOH, 10 g of Glauber's salt, 10 g of paraffin wax, 5 g of borax and stearic acid 5 g (A), 10 g (B) 15 g (C) and 20 g (D).

Activated carbon adsorbs water from salt and paraffin, which affects the reduction of vaporization energy, is significantly higher than the solubility of salt or paraffin. Activated carbon can affect the ratio of energy vaporization to the melting energy of the composite [3, 16]. As the mixture is being mixed and water is being added, a point is reached at which the water is sufficient to wet all the particles. Continued mixing compacts the powder, removing air and the compacted powder starts to agglomerate into lumps. Continued mixing resulted in these lumps becoming plastic. Excess water makes these lumps too soft for the forming process. Clays have been used as binders in carbon mixtures to impart strength to the carbon body formed there from. Typically, water contents of about 140% to about 180%, and more typically about 145% to about 160% based on the carbon and clay content impart good plasticity and handle ability to the mixture [27].

Active powder carbon is suitable in a mixture with water and clay, which can absorb gases at elevated temperatures close to 100 °C, thereby maintaining the volume of the same constant [10, 27].

Besides, the thermal conductivity in molten state was larger than that in solid state. It also indicated that the activated carbon was favorable to form carbonaceous layers, which created a physical protective barrier on the surface of the composites. The protective barrier could confine the transfer of flammable molecules to the gas phase, and thus the thermal stability of the composites was improved [7, 35].

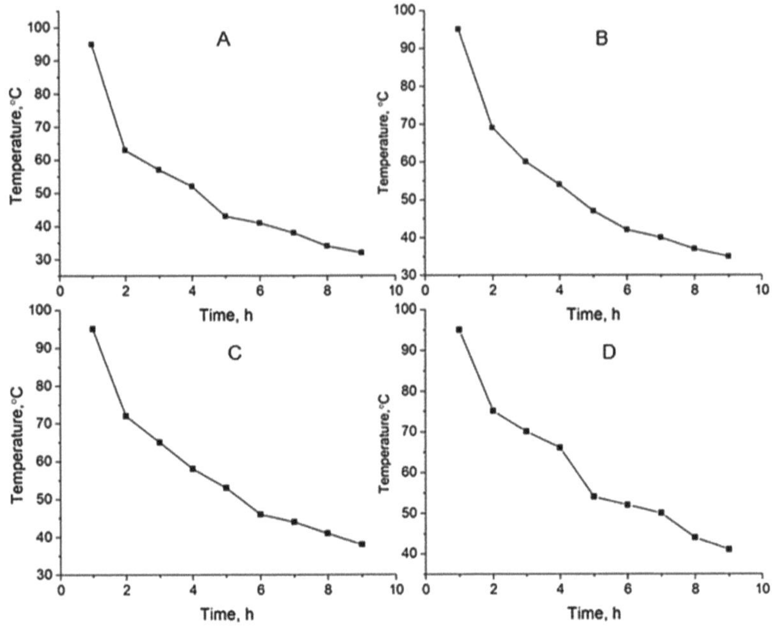

Fig. 4. Glauber's salt influence on temperature changes of heterogeneous gel. Heterogeneous gel composition: 100 g of water, 60 g of clay, 5 g of NaOH, 10 g of stearic acid, 10 g of paraffin wax, 5 g of borax and Glauber's salt 5 g (A), 10 g (B) 15 g (C) and 20 g (D).

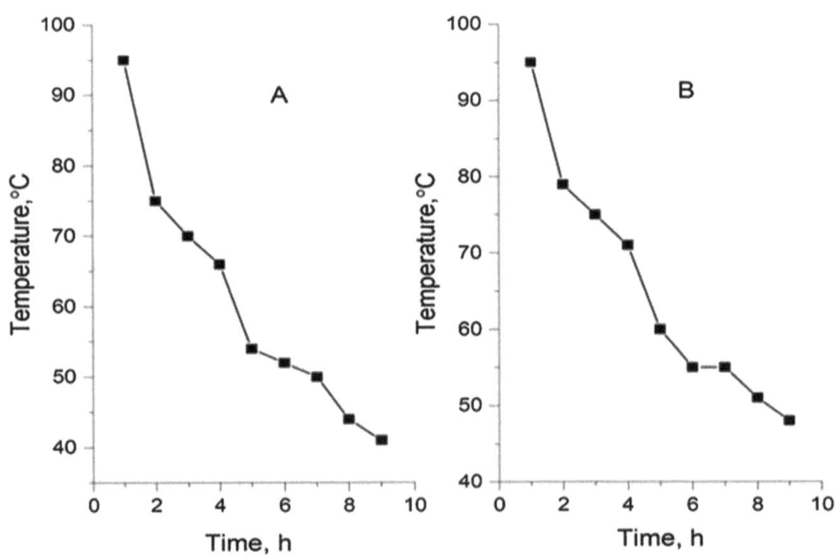

Fig. 5. Activated carbon influence on temperature changes in heterogeneous gel. Heterogeneous gel composition: (A) 100 g of water, 60 g of clay, 20 g of Glauber's salt, 5 g of NaOH, 10 g of stearic acid, 10 g of paraffin wax and 5 g of borax and; (B) 100 g of water, 60 g of clay, 20 g of Glauber's salt, 5 g of NaOH, 10 g of stearic acid, 10 g of paraffin wax, 5 g of borax, and 10 g of activated carbon.

For hydrocarbon adsorption applications, the preferred type of activated carbon is what is considered to be a collection of very small graphitic platelets which are bound together with an open structure leading to high surface area [5].

During phase changing, the solid-liquid interface moves away from the heat transfer surface, and the heat transfer resistance gradually increases due to the increased thickness of the molten/solidified medium. By introducing an expanded gel with bentonite clay, this effect is significantly reduced [36].

4 Conclusion

In this work, we managed to homogenize a heterogeneous mixture of inorganic and organic PCMs agents. In the homogeneous bentonite clay gel, 20–25% PCM was added. This gel contains about 50% water that significantly affects heat increase of the specific mixture. Obtained gel based on clay allows continuous temperature change and heat transfer, while Glauber's salt, stearic acid and paraffin caused the latent heat of phase at different temperature intervals.

Expanded gel based on bentonite clay is a carrier by which the addition of PCMs can be used to change thermodynamic properties. Composite based on expanded bentonite and clay and PCMs has significantly higher specific heat from the mixture of PCMs. Addition of activated carbon in gel provides additional system homogenization and increased accumulation of heat. In addition, activated carbon significantly reduced the cooling of the gel, which indicates further investigation of activated carbon concentration increase in the future. It is also possible to insert micronized particles of some metal or salt in order to increase the thermal capacity of the system and to allow faster heating and cooling of the commercially available material.

The present-day composite gel based clay, active carbon and stearic acid gel is very suitable for thermotherapy in medicine packaged in elastic polyethylene bags that will be heated in a microwave oven. Due to its heterogeneous composition and possible corrosive properties, this composite should be encapsulated in polymeric capsules, tubes, spheres or some other geometric shapes.

References

1. Stojiljković, S.T., Todorović, B.Ž.: The adsorption-desorption power of bentonite based materials. LAMBERT Academic Publishing, Saarbrücken (2016)
2. Sadek, O.M., Mekhemer, W.K.: Na-montmorillonite clay as thermal energy storage material. Thermochim. Acta **370**, 57–63 (2001)
3. Stojiljković, S., Savić, I., Mitković, P., Vasić, L., Marinković, A.: An urban planning approach to the climatization of space using natural resources based on ceramic clay, zeolite and bentonite clay. Sci. Sinter. **46**(2), 259–268 (2014)
4. Al-Kayiem, H.H., Lin, S.C., Lukmon, A.: Review on nanomaterials for thermal energy storage technologies. Nanosci. Nanotechnol. Asia **3**, 60–71 (2013)
5. Wuttig, M., Steimer, C.: Phase change materials: from material science to novel storage devices. Appl. Phys. A-Mater. Sci. Process. **87**(3), 411–417 (2007)

6. Stojiljkovic, S., Miljkovic, V., Nikolic, G., Kostic, D., Arsic, B., Barber, J., Savic, I., Savic, I.: The influence of the addition of polymers on the physico-chemical properties of bentonite suspensions. Sci. Sinter. **46**(1), 65–73.44 (2014)
7. Cao, L.: Properties evaluation and applications of thermal energy storage materials in buildings. Renew. Sustain. Energy Rev. **48**, 500–522 (2015)
8. Todorović, B.Ž., Stojiljković, S.T., Stojiljković, D.T., Petrović, S.M., Takić Lj, M., Stojiljković, M.S.: Removal of As^{3+} cations from water by activated carbon, bentonite and zeolite in a batch system at different pH. J. Elementol. **22**(2), 713–723 (2017)
9. Stojiljkovic, S.: Role of bentonite clay in the ecology of the human body. In: XIV International Conference on Medical Geology (GEOMED), pp. 20–25 (2011)
10. Veniale, F.: The role of microfabric in clay soil stability. Mineral. Petrogr. Acta **29**, 101–119 (1985)
11. Chen, Z., Shan, F., Cao, L., Fang, G.Y.: Synthesis and thermal properties of shape stabilized lauric acid/activated carbon composites as phase change materials for thermal energy storage. Solar Energy Mater Solar Cells **102**, 131–136 (2012)
12. Hamdan, M.A., Al-Hinti, I.: Analysis of heat transfer during the melting of a phase-change material. Appl. Therm. Eng. **24**(13), 1935–1944 (2004)
13. Li, Z.B.: Paraffin/diatomite/multi-wall carbon nanotubes composite phase change material tailormade for thermal energy storage cement-based composites. Energy **72**, 371–380 (2014)
14. Prekajski, M., Mirković, M., Todorović, B., Matković, A., Marinović-Cincović, M., Luković, J., Matović, B.: Ouzo effect – new simple nanoemulsion method for synthesis of strontium hydroxyapatite nanospheres. J. Eur. Ceram. Soc. **36**(5), 1293–1298 (2016)
15. Mondal, S.: Phase change materials for smart textiles – an overview. Appl. Therm. Eng. **28**, 1536–1550 (2008)
16. Nikolić, L., Ristić, I., Stojiljković, S., Vuković, Z., Stojiljković, D., Nikolić, V., Budinski-Simendić, J.: The influence of montmorillonite modification on the properties of composite material based on poly(methacrylic acid). J. Compos. Mater. **46**, 921–928 (2012)
17. Wang, S.X., Li, Y., Hu, J.Y., Tokura, H., Song, Q.W.: Effect of phase change material on energy consumption of intelligent thermal protective clothing. Polym. Test. **25**(5), 580587 (2006)
18. Li, Y.: The science of clothing comfort. Text. Prog. **31**, 1–135 (2001)
19. Shukla, N., Fallahi, A., Kosny, J.: Performance characterization of PCM impregnated gypsum board for building applications. Energy Procedia **30**, 370–379 (2012)
20. Iten, M., Liu, S., Shukla, A.: Renew. Sustain. Energy Rev. **61**, 175–186 (2016)
21. Sari, A.: Thermal energy storage characteristics of bentonite-based composite PCMs with enhanced thermal conductivity as novel thermal storage building materials. Energy Convers. Manag. **117**, 132141 (2016)
22. Sharma, A., Tyagi, V.V., Chen, C.R., Buddhi, D.: Review on thermal energy storage with phase change materials and applications. Renew. Sustain. Energy Rev. **13**(2), 318–345 (2009)
23. Maciej, J., Shady, A.: Thermal conductivity of gypsum with incorporated phase change material (PCM) for building applications. J. Power Technol. **91**(2), 49–53 (2011)
24. Zastawna-Rumin, A.: Phase change materials vs. internal temperature in a building. Techn. Trans. **111**(8-A), 207–213 (2014)
25. Pan, L., Tao, Q., Zhang, S., Wang, S., Zhang, J., Wang, S., Wang, Z., Zhang, Z.: Preparation, characterization and thermal properties of microencapsulated phase change materials. Sol. Energy Mater. Sol. Cells **98**, 6670 (2012)
26. Marín, J.M., Zalba, B., Cabeza, L.F., Mehling, H.: Improvement of a thermal energy storage using plates with paraffin–graphite composite. Int. J. Heat Mass Transfer **48**, 2561–2570 (2005)

27. Karaipekli, A., Sarı, A., Kaygusuz, K.: Thermal conductivity improvement of stearic acid using expanded graphite and carbon fiber for energy storage applications. Renew Eng. **32**, 2201–2210 (2007)
28. Memon, S.A., Cui, H.Z., Zhang, H., Xing, F.: Appl. Energy **139**, 43–55 (2015)
29. Cabeza, L.F., Barreneche, C., Martorell, I., Miró, L., Sari-Bey, Fois S.M., Paksoy, H.O., Sahan, N., Weber, R., Constantinescu, M., Anghel, E.M., Malikova, M., Krupa, I., Delgado, M., Dolado, P., Furmanski, P., Jaworski, M., Haussmann, T., Gschwander, S., Fernández, A.I.: Unconventional experimental technologies available for phase change materials (PCM) characterization. Part 1. Thermophysical properties. Renew. Sustain. Energy Rev. **43**, 1399–1414 (2015)
30. Gschwander, S.: Standardization of PCM characterization via DSC. In: The 13th international Conference on Energy Storage, Greenstock, Beijing (2015)
31. De Garcia, A., Cabeza, L.F.: Phase change materials and thermal energy storage for buildings. Energy Build. **104**, 414–419 (2015)
32. Stojiljkovic, S., Miljkovic, V., Nikolic, G., Kostic, D., Arsic, B., Barber, J., Savic, I.: The influence of the addition of polymers on the physico-chemical properties of bentonite suspensions. Sci. Sinter. **46**(1), 65–73 (2014)
33. Soares, N.M.L.: (2016) Thermal energy storage with phase change materials (PCMs) for the improvement of the energy performance of buildings. Tese de doutoramento, Coimbra. Disponível na. www: http://hdl.handle.net/10316/29306
34. Lamberg, P.: Approximate analytical model for two phase solidification problem in a finned phase-change-material storage. Appl. Energy **77**, 131–152 (2004)
35. Li, B.X., Liu, T.X., Hu, L.Y., Wang, Y.F., Nie, S.B.: Facile preparation and adjustable thermal property of stearic acid–graphene oxide composite as shapestabilized phase change material. Chem. Eng. J. **215**, 819–826 (2013)
36. Stojiljković, S., Savić, I., Mitković, P., Vasić, L., Marinković, A.: An urban planning approach to the climatization of space using natural resources based on ceramic clay, zeolite and bentonite clay. Sci. Sinter. **46**(2), 259–268 (2014)

Experimental Investigation of Mechanical Properties on Friction Stir Welded Aluminum 2024 Alloy

Miodrag Milcic[1(✉)], Tomaz Vuherer[2], Igor Radisavljevic[3], and Dragan Milcic[1]

[1] Faculty of Mechanical Engineering, University of Nis, Aleksandra Medvedeva 14, 18000 Nis, Serbia
miodrag.milcic@masfak.ni.ac.rs
[2] Faculty of Mechanical Engineering, University of Maribor, Smetanova ulica 17, 2000 Maribor, Slovenia
[3] Military Technical Institute, Ratka Resanovica 1, 11000 Belgrade, Republic of Serbia

Abstract. The paper presents the results of structural and mechanical testing of the alloyed aluminum alloys AA 2024 welded by the FSW process. Investigations were conducted on the welding machine, built on the base of the conventional, vertical milling machine. The goal of the research was to know the relationship between welding parameters and mechanical and microstructural properties of 2024 joints. The following welding parameters were used: the rotation speed of the tool did not change and amounted to 750 rpm, and the welding speed was 73, 116 and 150 mm/min. The welded joints obtained were free of errors and with an acceptable flat surface.

Keywords: Friction stir welding · Aluminum alloy 2024
Microstructural properties · Mechanical properties

1 Introduction

Aluminum alloys have been widely used in automotive and aerospace industries. Both industries are pushing the boundaries of new innovative products, a requirement for greater capacity and, at the same time, a lower weight with a robust design. Aluminum alloys are characterized by high load capacity relative to the mass level at a relatively low price. There are no good anti-corrosive properties and as a rule, they are poorly welded by conventional welding procedures.

Welding is a fabrication process used to join materials, usually metals or thermoplastics, together. During welding, the pieces to be joined (the workpieces) are melted at the joining interface and usually, a filler material is added to form a pool of molten material (the weld pool) that solidifies to become a strong joint. For the welding of aluminum parts, can be used the welding processes of GMAW (MIG) and GTAW (TIG) and Friction Stir Welding (FSW) as a solid-state joining technique. Thus, FSW is a very suitable, and increasingly used, for joining high strength aluminum alloys (2xxx,

N. Mitrovic et al. (Eds.): CNNTech 2018, LNNS 54, pp. 44–58, 2019.
https://doi.org/10.1007/978-3-319-99620-2_4

6xxx, 7xxx and 8xxx series), that currently applied to the aerospace, automotive, marine and military industries.

The original application for friction stir welding was the welding of long lengths of the materials in the aerospace, shipbuilding, and railway industries. Examples include large fuel tanks for space launch vehicles, cargo decks for high-speed ferries, and roofs for railway carriages. In the last several years, the automotive industry has been aggressively studying the application of FSW in its environment. The drive to build more fuel-efficient vehicles has led to the increased use of aluminum in an effort to save on weight, which also improves recyclability when the vehicles are scrapped.

FSW was invented at The Welding Institute (TWI) of UK in 1991 as a solid-state joining technique, and it was initially applied to aluminum alloys [1]. The first (national) standard for friction stir welding of Aluminum Alloys for Aerospace Applications was adopted in December 2009 in the United States, and the first international standards are adopted in December 2011 [8, 13].

Tool geometry plays an important role in the FSW process. An FSW tool has two basic functions:

- the friction heat generated at the interface between the tool shoulder and the workpiece and
- the material plastic deformation heat that generated near the tool pin.

The friction heat input from the tool shoulder is believed to be the main heat input in the FSW process [2].

The amount of heat generated in the FSW process depends on the following parameters:

- technological process parameters: tool rotation speed n [rpm], speed tool movements v [mm/min], duration of the welding process, tool geometry, loads (axial force F_z, torque M_t etc.), etc.,
- tribological parameters: contact surface, contact pressure p, friction coefficient μ, roughness, wear, transfer of mass, a hardness of the body in contact, etc.),
- thermomechanical properties of materials: strength, modules elasticity, stiffness, deformability, etc. and
- other parameters: electrical and magnetic properties of materials, weldability, impurities, imperfections in material, structure material, suspension system, operating system propulsion system, etc.

The research was carried out at the Faculty of Mechanical Engineering in Nis. The process of generating heat in the FSW procedure and the development of an analytical model for determining the amount of generated heat was given in [14–20].

Thermo-mechanical simulation of friction stir welding can predict the transient temperature field, the active stresses developed, and the forces in all three dimensions and may be extended to determine the residual stress. The thermal stresses constitute a major portion of the total stress developed during the process. The numerical code developed at the Faculty of Mechanical Engineering in Nis, is a synergy of experimental models, analytical models, and numerical calculations. Numerical simulation of FSW included well known finite difference method for numerical estimation of temperatures in discrete nodes of workpieces and accuracy of the simulation is improved by the innovative numerical method for material flow definition – node substitution and replacements [7, 12, 20].

There has been a number of reports [2–11] highlighting the microstructural changes due to plastic deformation and frictional heat associated with FSW. Mechanical failure of the welds can take place in the SZ, TMAZ, or HAZ region depending on the amount of energy input which is controlled by the welding parameters such as rotational and travel speed. Since the material flow behavior is predominantly influenced by the material properties such as yield strength, ductility and hardness of the base metal, tool design, and FSW process parameters, the dependence of weld microstructure on process parameters differs in different aluminum alloys for a given tool design. In [24] is shown that the microstructure-based simulation is a reasonable tool to study the deformation behavior of FSW materials, which is difficult to be predicted within macroscopic models alone.

By knowing the parameters of welding and their change, the amount of energy input and the level of heating of the welded pieces are changed. In this way, using the properly selected welding parameters, the optimum state and flow of the materials of the welding pieces are achieved, necessary for the proper unwinding of the coupling process and obtaining the joint of the required quality. Authors of publication [9, 10, 21] have reported a quantitative investigation of the effect of parametric friction stir welding on the mechanical, structural and corrosion properties of the aluminum alloys.

The main goal of the extensive experimental research is the analysis of the influence of FSW parameters on the structural, mechanical properties and fatigue properties of the FSW butt joint of the aluminum alloy EN AW 2024-T351.

2 Experimental Work

The experiment was aimed to find the influence of input kinematic parameters such as welding speed (v) and tool rotation speed (n) on metallurgical and mechanical characteristics of welded joints. The base material was aluminum alloy EN AW 2024-T351.

The chemical composition of experimental plates is provided in Table 1 and mechanical properties are given in Table 2.

Table 1. Chemical composition of AA 2024-T351

Chemical composition	Cu	Mg	Mn	Fe	Si	Zn	Ti
wt%	4.70	1.56	0.65	0.17	0.046	0.11	0.032

The dimensions of welded plates were 500 mm × 65 mm × 6 mm. Both sides of the welding plates are machined on the grinder at a thickness of 6 mm. Before the start of welding, an austenitic plate is placed under the welding plates as a backing plate. A milling machine was used for welding. The weld length was approximately 400 mm.

Table 2. Mechanical properties of AA 2024-T351

Yield strength R_{eh} (MPa)	Ultimate tensile strength R_m (MPa)	Elongation A_5 (%)	Hardness HV
370	481	17.9	137

Figure 1 shows a machine and Fig. 2 presents a tool used for butt joint FSW.

Fig. 1. Machine for FSW welding

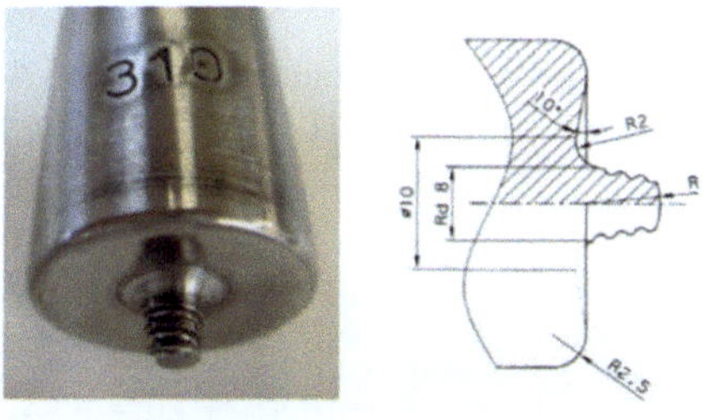

Fig. 2. Fabricated FSW tools

Welding was made in accordance with the planning matrix of the experiment, with variations in tool rotation speed (n) and welding speed (v), Table 3. Other parameters of welding were maintained constant.

Table 3. Friction stir welding parameters

Sample	Rotation speed n (rpm)	Welding speed v (mm/min)	Ratio n/v (rev/mm)	Ratio v/n (mm/rev)
A-I	750	73	10.27	0.0973
B-II		116	6.47	0.154
C-III		150	5	0.2

After the welding process was completed, welds were tested. For that purpose, visual control was performed, on the weld face and root of the seam, as well as the radiographic control of samples. No defects were detected (visually, touch or magnifier).

Appearances of upper and bottom butt FSW joint surface are shown in Fig. 3.

Fig. 3. Butt FSW joint

Specimens are made from FSW welded samples. Figure 4 shows the part of specimens that are used for tensile testing, impact strength testing and SENB specimens.

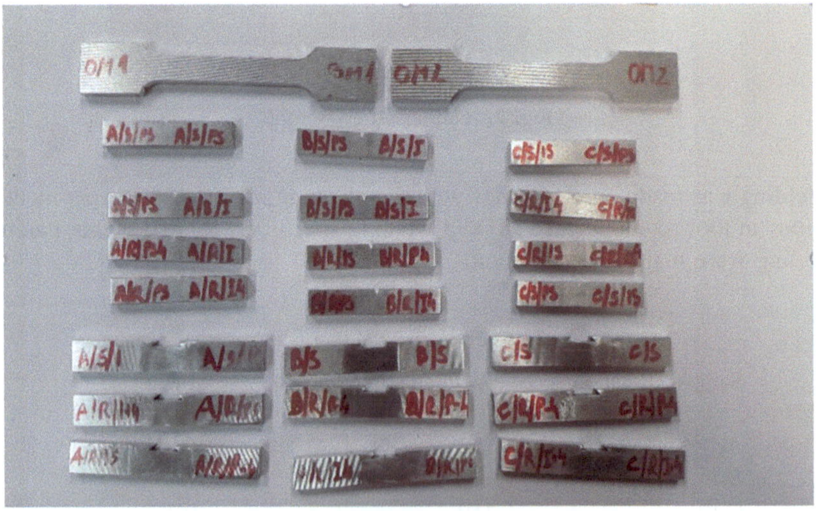

Fig. 4. Tensile specimens, impact test specimens and SENB specimens

The dimensions of the test specimens for tensile testing are shown in Fig. 5. To evaluate the impact behavior of the friction stir weldments, the Charpy impact test notched specimens (CVN) are machined. Specimen size is 10 × 6 × 55 mm with 45° V-notch of 2 mm depth and 0.25 mm root radius (Fig. 6). These are reduced size non-standard specimens whose dimensions are dictated by the thickness of the plate. Notches are machined at the rear side of the welding direction, precisely along the welding line of weldments.

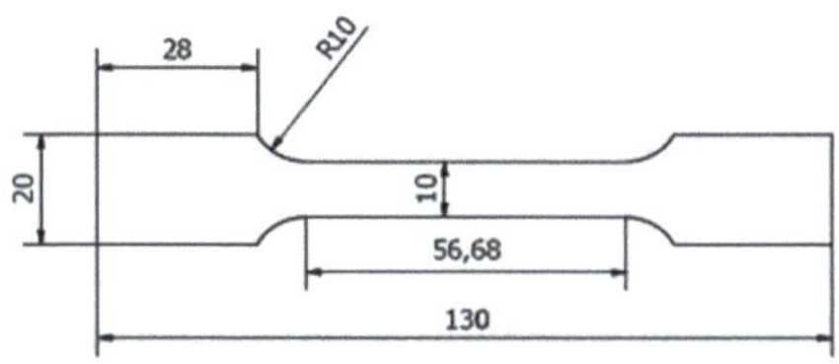

Fig. 5. Dimension of tensile specimens

The metallographic observation was carried out by optical microscopy (OM) using Leica M205A optical microscope. The specimen for OM was ground, polished and etched using Tucker's (45 ml HCl, 15 ml HNO3, 5 ml HF and 25 ml H2O) reagent. Much care was taken to ensure location-to-location correspondence between the structural observations and hardness measurements. The nugget zone average size measurements were processed using Leica DFC295 camera and LAS software.

Room-temperature tensile tests were carried out at a strain rate of $3.3 \times 10^{-3}\ s^{-1}$ on ASTM E8M transverse tensile specimens (Fig. 6). In order to assess the reproducibility, 2 specimens were tested and the average value was reported. Bend testing was carried out according to EN 910 with joint centered over the mandrel. The bending specimens were tested using face and root side of the joint in tension.

Vickers hardness measurement was conducted perpendicular to the welding direction, a cross section of weld joint, using digitally controlled hardness test machine (HVS-1000) applying 9.807 N force for 15 s. Hardness profiles were obtained along 3 horizontal (face joint, mid joint and root joint) and 63 vertical lines at a distance of 0.5 mm. Figure 9 shows hardness distribution across the welded joint at a different applied rotation speed (n) and welding speed (v).

Charpy impact tests are carried out at room temperature using instrumented impact pendulum. The pendulum is equipped with load cells positioned on a striker edge. The measuring device is connected to high-speed data acquisition with response time in terms of milliseconds. The instrumented system enabled to collect instantaneous load and time data from pendulum during the "fracture opening time".

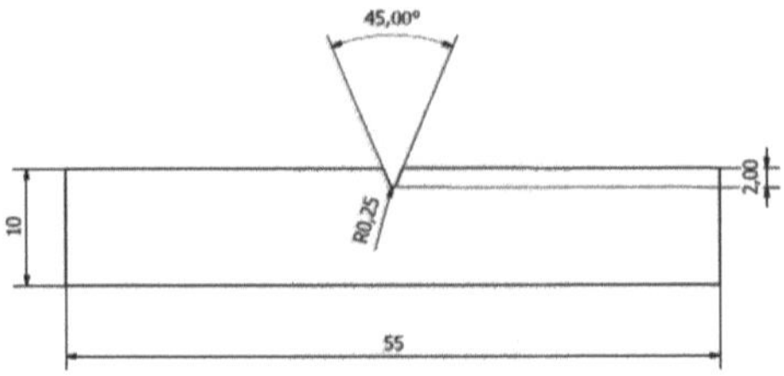

Fig. 6. Dimension of Charpy impact test specimens

3 Results and Discussions

Figure 7 shows macroscopic appearances of the cross-section of joints welded at 750 min^{-1} utilizing three welding speed (73, 116 and 150 mm/min). All welded joints have the porosity as defects. HAZs are recognized as slightly dark areas on both sides of SZ. Average grain size in SZ depends mainly on the peak temperature and strain rate. The grain size in SZ increases with increasing peak temperature and decreasing strain rate. Since all the welds were fabricated at the constant rotation speed in the present study, strain rate is constant and therefore, grain size in SZ depends only on the peak temperature.

Generally, the temperature in welding material increases with decreasing the welding speed and increasing the tool rotation speed. This means that the minimum amount of heat generated for the welding parameters is 750/150 (C-III), and the largest amount of heat for the welding parameters is 750/73 (A-I). Figure 8 shows that the grain sizes in SZ of (C-III) were lower than for others welding parameters.

Figures 9, 10 and 11 show the microhardness profiles along the line near the face of the weld, in the middle of the thickness of the parts welded and near the root of the welds for welded samples with different welding parameters.

The average hardness value in the nugget and in the base material is similar among all samples for position Face joint. In almost all tests, the highest value of micro-hardness in the stirred zone was found not in the middle of the joint but shifted towards one side of the joint, where the higher plastic strain was observed and the micro-hardness curve shows a W-shape.

The hardness distribution for the Sample A-I obtained by welding with the FSW parameters 750/73 in the zone of the nugget, over the entire height of the weld is approximately equal to the hardness. For Sample C-III (750/150), with the smallest heat generated, the hardness distribution at the Mid joint and Root joint level is lower compared to Face joint. At about 18 mm from the weld centerline, hardness starts to decrease and reaches a minimum value of about 102 HV at 12.5 mm from the weld center line. The minimum points are located on both sides of HAZ. As coming to the weld centerline through SZ beyond that minimum point, hardness turns to increase and reaches again equal value to that of BM.

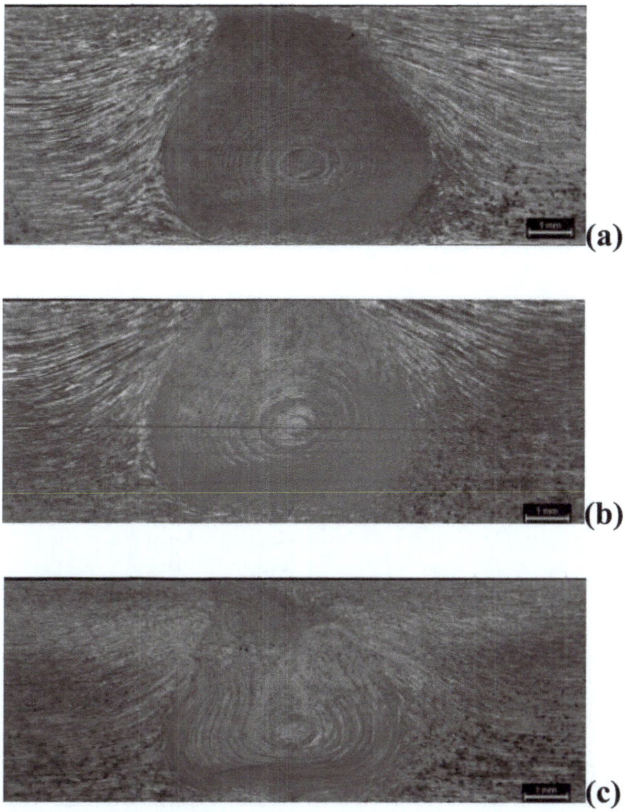

Fig. 7. Macrostructures of the transverse cross section at constant rotation speed 750 min^{-1}: (a) A-I, (b) B-II, and (c) C-III

Tensile testing was performed for all tree FSW joints. The tensile testing results in FSW joints are given in Table 4. Among the tree FSW parameters studied, i.e., at 750/73, 750/116 and 750/150 rpm/(mm/min), the average tensile yield strength and ultimate tensile strength.

This variation of tensile strength with rotational speeds for a given traverse speed appears to be linked to the energy of the welds. Joint efficiency as high as 97% of base metal could be achieved at 750/116 rpm/(mm/min). The highest elongation of the welded joint is achieved with the welding parameters 750/116 rpm/(mm/min) and is 7.43%.

The testing of welded joints was also performed on bending, around the face, and around the root. The three-point bending test results FSW joints are given in Table 5. The welded FSW joint has poor bending characteristics. Comparing the obtained bending test results, the largest bend angle to the first cracking phenomenon is for welding parameters 750/116 rpm/(mm/min) and amounts to 42°.

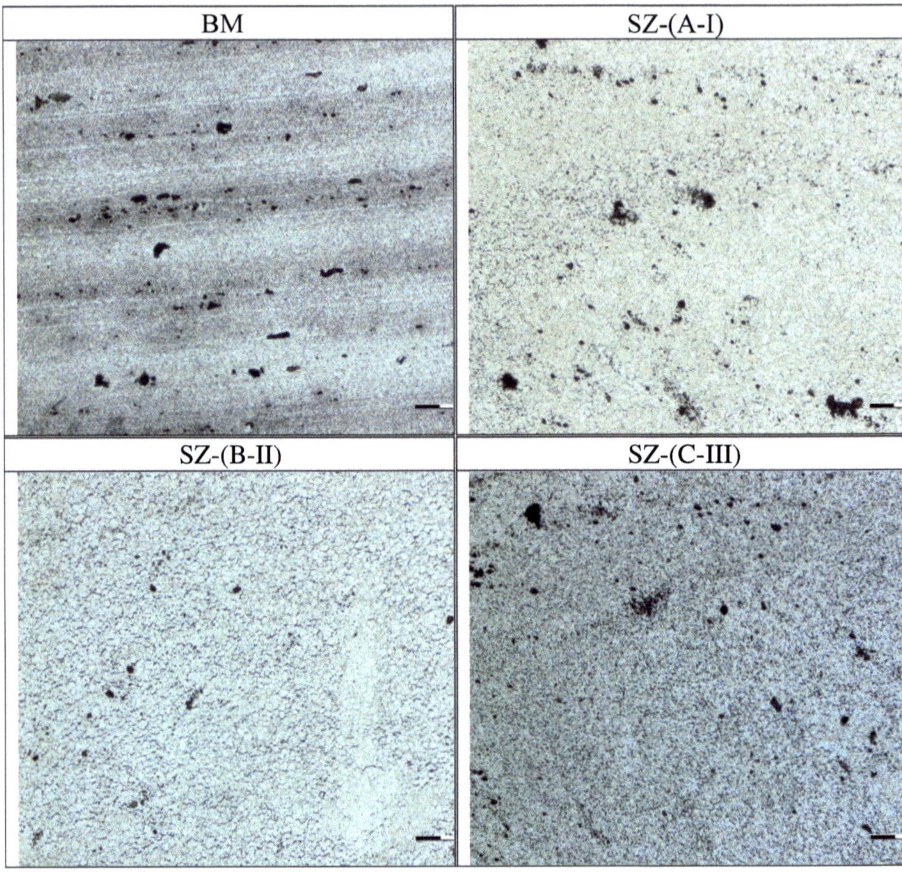

Fig. 8. Optical microstructures of BM and stir zone (SZ) at the same welding speed of 73, 116 and 150 mm/min

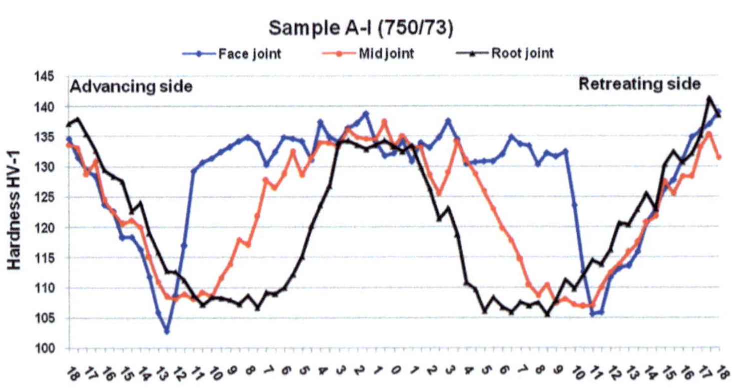

Fig. 9. Hardness profiles at a face, mid and root of FSW joint for welding parameters 750/73 rpm/(mm/min).

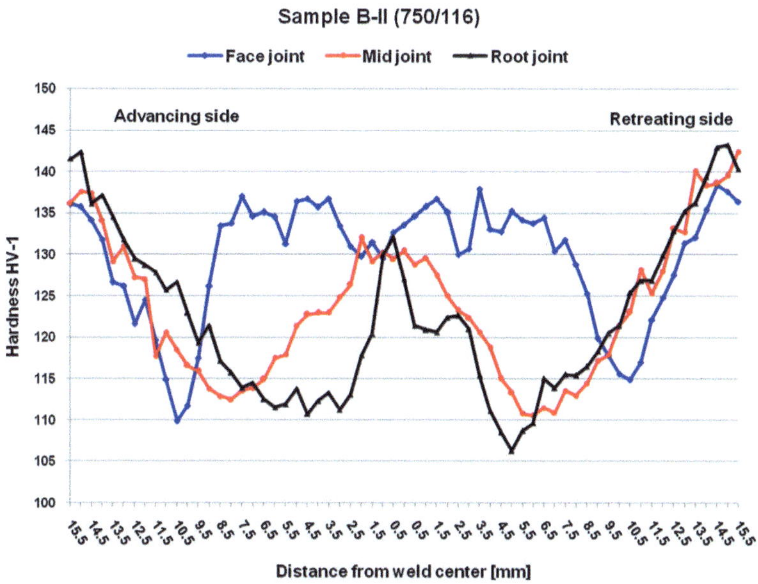

Fig. 10. Hardness profiles at a face, mid and root of FSW joint for welding parameters 750/116 rpm/(mm/min).

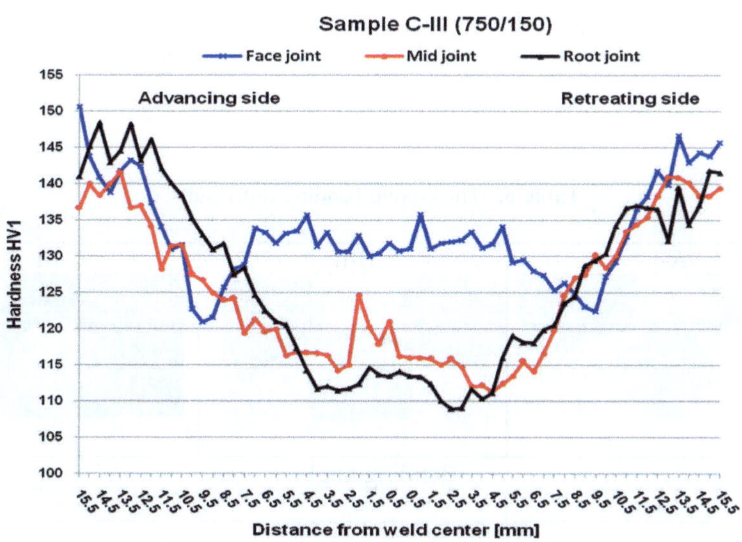

Fig. 11. Hardness profiles at a face, mid and root of FSW joint for welding parameters 750/150 rpm/(mm/min).

The Charpy impact tests result FSW joints are given in Table 6. The value of impact strength increases with welding speed from 73 mm/min to 116 mm/min. But the value of impact strength slightly decreases as the welding speed increases from 116 mm/min to 150 mm/min (Fig. 12) coincide with the experimental measurement results [23].

The best characteristics of impact toughness have a welded pattern with welding parameters 750/116 rpm/(mm/min) and an impact energy of 8.34 J and an impact toughness of 20.85 J/cm^2.

Table 4. Tensile testing results

A-I	B-II	C-III
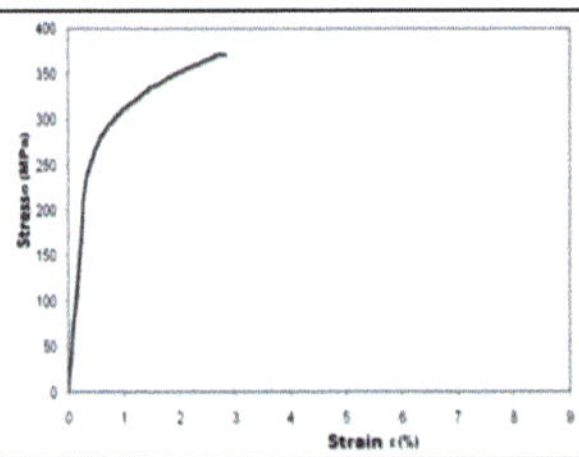	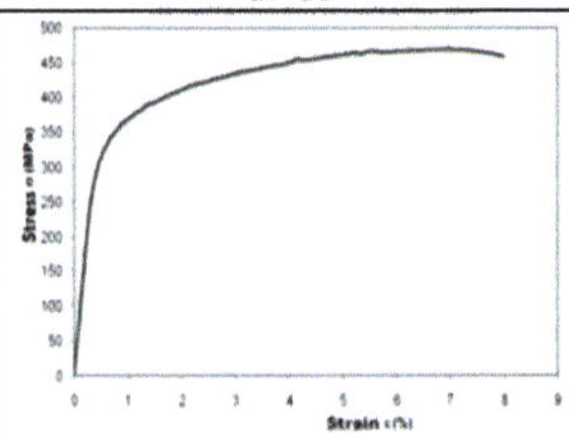	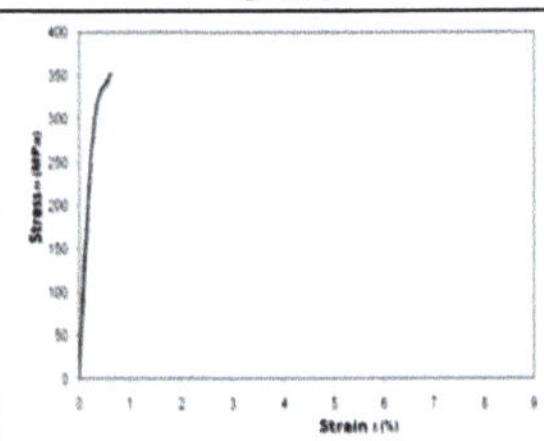
$R_{p0,2}$=281.9 MPa	$R_{p0,2}$=330.9 MPa	$R_{p0,2}$=337.6 MPa
Ultimate Tensile Strength (UTS) UTS=371.00 MPa	UTS=469.06 MPa	UTS=352.03 MPa
Elongation =2.29%	Elongation =7.43%	Elongation =0.33%
Joint efficiency=77%	Joint efficiency=97%	Joint efficiency=73%

Table 5. Three-point bending test results

A-I	B-II	C-III
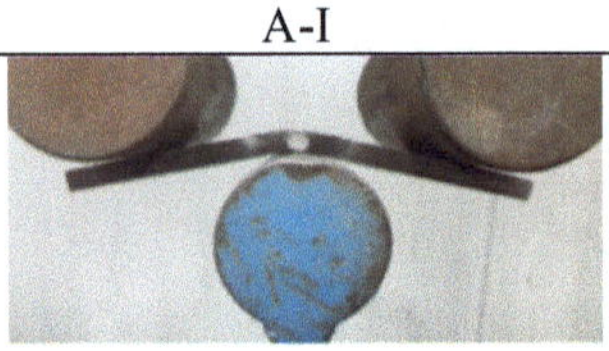	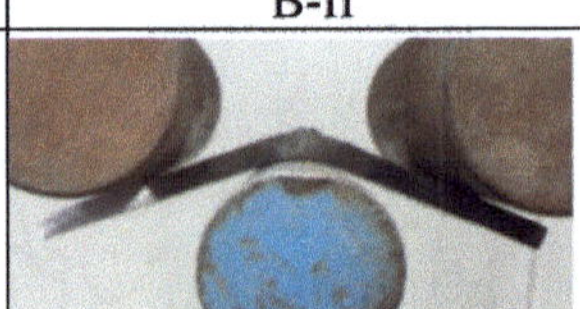	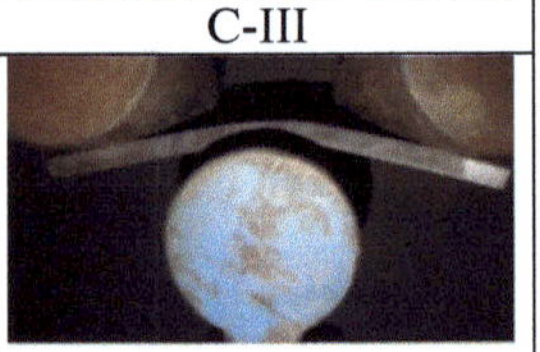
Bending angle		
≈24°	≈42°	≈26°

Table 6. Charpy impact tests results

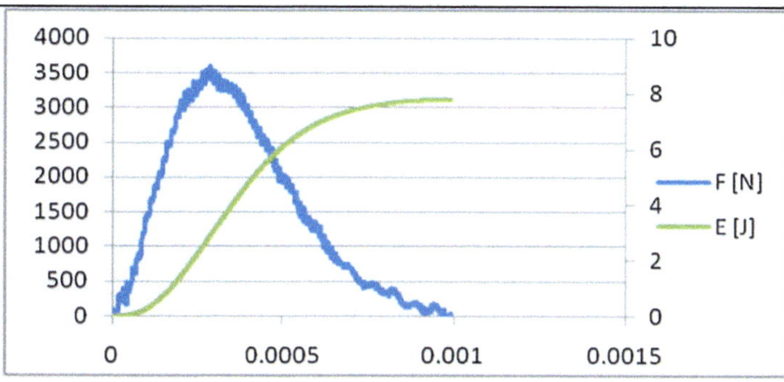

A-I
E=7.8 J
e=19.5 J/cm^2
Ei=2.88 J
Ep=4.91 J
UTS=371 MPa

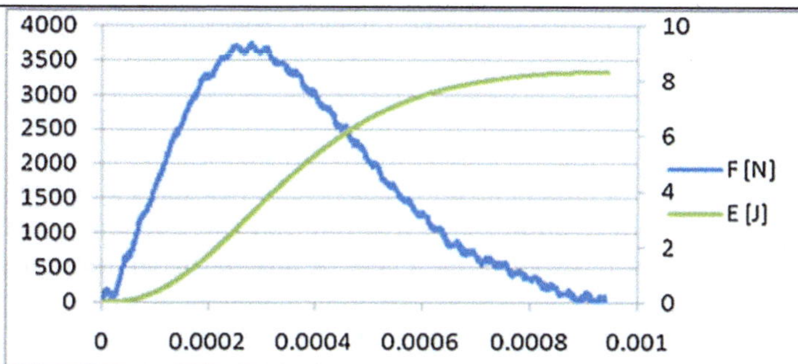

B-II
E=8.34 J
e=20.85 J/cm^2
Ei=3.19 J
Ep=5.15 J
UTS=469 MPa

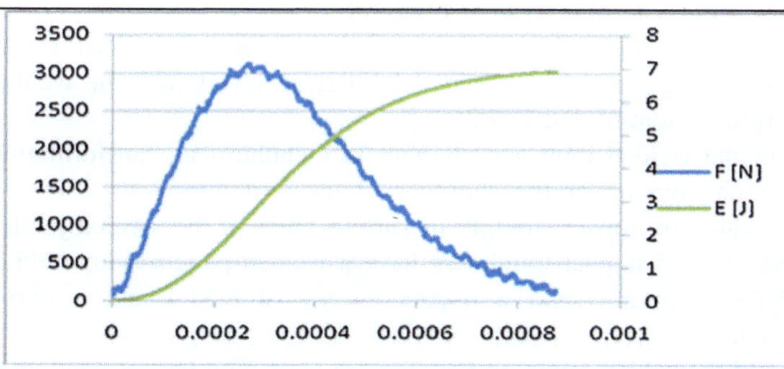

C-III
E=6.91 J
e=17.28 J/cm^2
Ei=0 J
Ep=6.91 J
UTS=352 MPa

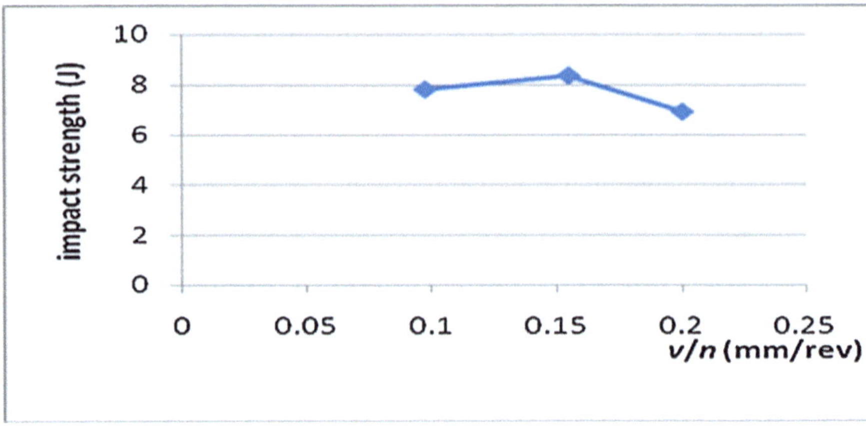

Fig. 12. Effects of process parameters on impact strength

4 Conclusions

In this study, friction stir welding of aluminum alloy 2024-T351 was studied by using a vertical milling machine with tool rotation speed n = 750 rpm and different welding speeds 73, 116 and 150 mm/min to evaluate the effect of process parameters on the mechanical properties.

The minimum frictional heat generated is for the welding parameters is 750/150 rpm/(mm/min) (C-III), and the largest amount of heat for the welding parameters is 750/73 rpm/(mm/min) (A-I). Grain size in SZ is the smallest in the case of the highest welding speed of 750/150 rpm/(mm/min).

Joint efficiency as high as 97% of base metal could be achieved at 750/116 rpm/(mm/min). The highest elongation of the welded joint is achieved with the welding parameters 750/116 rpm/(mm/min) and is 7.2%.

The asymmetry of the welded joint and changes in metallurgical transformations occurring around the pin and under the shoulder of the tool during its combined moving influence the value of impact strength in various areas of the welded joint.

The best characteristics of impact toughness have a welded pattern with welding parameters 750/116 rpm/(mm/min) and a impact energy of 8.34 J and a impact toughness of 20.85 J/cm^2.

The properties of FSW welded joints on bending are poor. The largest bend angle to the first cracking phenomenon is for welding parameters 750/116 rpm/(mm/min) and amounts to 42°.

References

1. Thomas, W.M., Nicholas, E.D., Needham, J.C., Murch, M.G., Temple-Smith, P., Dawes, C.J.: Friction stir butt welding. GB Patent No. 9125978.8, International Patent Application No. PCT/GB92/02203 (1991)
2. Ma, Z.Y., Mishra, R.S.: Friction stir welding and processing. Mater. Sci. Eng. **1**(1), 1–78 (2005)
3. Su, J.Q., Nelson, T.W., Mishra, R., Mahoney, M.: Microstructural investigation of friction stir welded 7050-T651 alloy. Acta Mater. **51**(3), 713–729 (2003)
4. Zimmer, S., Langlois, L., Laye, J., Bigot, R.: Experimental investigation of the influence of the FSW plunge processing parameters on the maximum generated force and torque. Int. J. Adv. Manuf. Technol. **47**, 201–215 (2010)
5. Gupta, R.K., Das, H., Pal, T.K.: Influence of processing parameters on induced energy, mechanical and corrosion properties of FSW butt joint of 7475 AA. JMEPEG **21**, 1645–1654 (2012)
6. Hussain, K.: Evaluation of parameters of friction stir welding for aluminium AA6351 alloy. Int. J. Eng. Sci. Technol. **2**(10), 5977–5984 (2010)
7. Dialami, N., Cervera, M., Chiumenti, M.: Numerical modelling of microstructure evolution in friction stir welding (FSW). Metals **8**(183), 1–15 (2018). https://doi.org/10.3390/met8030183
8. AWS D17.3/D17.3 M: 200x: Specification for Friction Stir Welding of Aluminium Alloys for Aerospace Applications, p. 60. An American National Standard, American Welding Society, Miami (2010)
9. Radisavljević, I., Živković, A., Grabulov, V., Radović, N.: Influence of pin geometry on mechanical and structural properties of butt friction stir welded 2024-T351 aluminum alloy. Hemijska Industrija **69**(3), 323–330 (2015)
10. Perović, M., Baloš, S., Kozak, D., Bajić, D., Vuherer, T.: Influence of kinematic factors of friction stir welding on the characteristics of welded joints of forged plates made of EN AW 7049 a aluminium alloy. Tech. Gazette **24**(3), 723–728 (2017)
11. Vilaça, P., Mendes, J., Nascimento, F., Quintino, L.: Application of FSW to join aluminium foil winding coils for electrical transformers. Int. J. Mech. Syst. Eng. **2**(2016), 115 (2016). https://doi.org/10.15344/2455-7412/2016/115
12. Mijajlović, M., Milčić, D., Milčić, M.: Numerical simulation of friction stir welding. Therm. Sci. **18**(3), 967–978 (2014)
13. Mijajlović, M., Milčić, D., Đurđanović, M., Grabulov, V., Živković, A., Perović, M.: Osnovni pojmovi kod postupka zavarivanja trenjem sa mešanjem prema AWS D17.3/D17.3M: 2010 i ISO 25239-1: 2011. Zavarivanje i zavarene konstrukcije, No. **2**, 61–68 (2012)
14. Mijajlović, M., Stamenković, D., Milčić, D., Đurđanović, M.: Study about friction coefficient estimation in friction stir welding. In: Proceedings of the 7th International Conference on Tribology, Balkantrib, vol. 11, pp. 323–330 (2011)
15. Mijajlović, M., Milčić, D., Anđelković, B., Vukićević, M., Bjelić, M.: Mathematical model for analytical estimation of generated heat during friction stir welding. Part 1. J. Balk. Tribol. Assoc. **17**(2), 179–191 (2011)
16. Mijajlović, M., Milčić, D., Anđelković, B., Vukićević, M., Bjelić, M.: Mathematical model for analytical estimation of generated heat during friction stir welding. Part 2. J. Balk. Tribol. Assoc. **17**(3), 361–370 (2011)

17. Mijajlović, M., Milčić, D., Stamenković, D., Živković, A.: Mathematical model for generated heat estimation during plunging phase of FSW process. Trans. Famena Faculty Mech. Eng, Naval Archit. Zagreb, Croatia **XXXV**(1), 39–54 (2011)
18. Milčić, D., Mijajlović, M., Pavlović, T.N., Vukić, M., Mančić, D.: Temperature based validation of the analytical model for the estimation of the amount of heat generated during friction stir welding. Therm. Sci. **16**(Suppl. 2), 337–350 (2012)
19. Mijajlović, M., Pavlović, T.N., Jovanović, S., Jovanović, D., Milčić, M.: Experimental studies of parameters affecting the heat generation in friction stir welding process. Therm. Sci. **16**(Suppl. 2), 351–362 (2012)
20. Mijajlović, M., Milčić, D.: Analytical model for estimating the amount of heat generated during friction stir welding: application on plates made of aluminium alloy 2024 T351, welding processes. In: Kovacevic, R. (ed.) In Tech. https://doi.org/10.5772/53563 (2012)
21. Radisavljevic, I., Zivkovic, A., Radovic, N., Grabulov, V.: Influence of FSW parameters on formation quality and mechanical properties of Al 2024-T351 butt welded joints. Trans. Nonferrous Metals Soc. China **23**(12), 3525–3539 (2013)
22. Živković, A., Mijajlović, M., Daşcău, H., Sedmak, A., Radisavljević, I., Milčić, D., Veljić, D.: Influence of the welding tool's geometry on productivity of friction stir welding process. In: The 5th International Conference Innovative Technologies for Joining Advanced Materials, Timisoara, Romania, Tima, 16–17 June 2011, vol. 11, pp. 1–4 (2011)
23. Sudhir, K., Pardeep, K.: Effect of parameters of friction stir welding on the impact strength of aluminium 6063. Int. J. Curr. Eng. Technol. **6**(3), 993–998 (2016)
24. Balokhonov, R., Romanova, V., Batukhtina, E., Sergeev, M., Emelianova, E.: A numerical study of the microscale plastic strain localization in friction stir weld zones. Facta Univ. Ser. Mech. Eng. **16**(1), 77–86 (2018)

Possibility for Removing Products of Thermal Degradation of Edible Oil by Natural Aluminosilicates

Sanja Dobrnjac[1(✉)], Mirko Dobrnjac[2], Jelena Penavin Skundric[3], Ljubica Vasiljevic[4], Stevan Blagojevic[5], and Zvjezdana Sandic[6]

[1] Projekt JSC, Veselina Maslese St. 1/IV, 78000 Banjaluka, Bosnia and Herzegovina
sanjadob@gmail.com
[2] Faculty of Mechanical Engineering, University of Banjaluka, Stepe Stepanovica St. 73, 78000 Banjaluka, Bosnia and Herzegovina
[3] Faculty of Natural Sciences and Mathematics, University of Banjaluka, Mladena Stojanovica St. 2, 78000 Banjaluka, Bosnia and Herzegovina
[4] Faculty of Technology, University of East Sarajevo, Karakaj, 75400 Zvornik, Bosnia and Herzegovina
[5] Institute of General and Physical Chemistry, Studentski trg 12/V, 11000 Belgrade, Serbia
[6] Faculty of Natural Sciences and Mathematics, Mladena Stojanovica St. 2, 78000 Banjaluka, Bosnia and Herzegovina

Abstract. Everyday food preparation often involves cooking on vegetable oil, where heating at temperatures over 110 °C changes the quality of oil due to various chemical reactions. In this paper, the degree of thermal degradation of edible oil was investigated when heating oil at a temperature ranging from 110 to 190 °C for 10 and 30 min. The reference to the quality of edible oil is the content of free fatty acids (FFA) which increases as the temperature rises, due to hydrolysis.

By removing the thermal degradation products of the edible oil used, it would be possible to prolong the time of use of the oil for other useful purposes, and thus its final disposal as waste, would be postponed, which is in accordance with environmental protection requirements. In line with this, the possibility of adsorption of FFA on natural aluminosilicates was investigated. The results obtained show a decrease in FFA content of 72 to 80%, depending on the time of heating and the type of adsorbent used. Further research should be continued on the regeneration of used edible oil by testing other quality parameters, as well as the application of other adsorbents

Keywords: Edible oil · Thermal degradation · Free fatty acids (FFA) Adsorption

1 Introduction

Refined edible sunflower oil is the most commonly used oil for frying food in our region. By heating oil above 150 °C, thermoxidation changes and the chemical reactions of hydrolysis, oxidation and polymerization take place, with increasing free fatty

N. Mitrovic et al. (Eds.): CNNTech 2018, LNNS 54, pp. 59–71, 2019.
https://doi.org/10.1007/978-3-319-99620-2_5

acids (FFA), non-volatile substances, peroxide number, and decreasing saponification and iodine number [1, 2]. The viscosity of the oil increases by increasing the polymeric products in the oil after frying. Also, the nutritive value of the oil, as well as the sensory and functional characteristics, is changing. The oil is degraded by temperature and oxidation [3, 4]. The intensity of the thermoxidation changes depends on the type of oil, temperature and period of heating. Sunflower oil, most commonly used in our area, is unfavorable from the aspect of heating, because of the high content of linoleic fatty acid [5–7].

The contents of moisture, FFA, non-volatile substances and peroxide number are important indicator to illustrate the quality as well as purity of the edible oil [8]. Changing the oil quality due to heating requires the end of oil use and its disposal in some of the regulated ways. Larger consumers of edible oil have a legal obligation to deliver used oil to authorized companies and store it in environmentally friendly way. Although the Waste Management Law promotes the regeneration and reuse of materials used in a single life cycle, edible oils collected from larger consumers most often finish as an energy source. However, large quantities of used oil, from public and household use, through sewage systems finish in watercourses, where it contaminates large amounts of water.

Surface and groundwater pollution by different waste is widespread in highly industrialized and urbanized cities due to direct discharge of such effluents into watercourses [9]. Among all pollutants in waste water, oil or fat is one of the most complicated substances due to the characteristics of these waste water that contain a very high amount of oil-in-water emulsions, as their basic contaminant. These emulsified oil droplets are sheltered from spontaneous coalescence into larger flocs, thus making oil separation by simple gravity a difficult and time consuming process.

Water pollution with edible oil can be irreversible for aquatic organisms, and the consequences of these effects are transferred, indirectly or directly, to humans, as they are also involved in the food chain of the ecosystem. The presence of oil and grease in water leads to the formation of oil layer, which causes significant pollution problem such as reduction passing of light and photosynthesis. It further complicates oxygen transfer from atmosphere to water and this leads to decreased amount of dissolved oxygen at the bottom of the water and this adversely affects survival of aquatic life in water [10]. In addition, the presence of oil and grease in wastewater include physical blockages in sewage system, pumps and other equipment, and these consequently lead to increase maintenance costs [11]. Furthermore, oil and grease interfere with aerobic biological wastewater treatment processes by reducing oxygen transfer rates. Equally, oil and grease reduce the efficiency of anaerobic treatment processes by reducing the transport of soluble substrates [12, 13]. Similarly, in the municipal water treatment plants, oil and grease cause unpleasant odors, turbidity and film, and make filtration treatment more difficult [14].

In order to protect the waterways, soil and human health, it would be useful to find a solution to extend the use of edible oil, to minimize the amount of oil in the environment as waste. Therefore, this paper examines the changes of oil quality parameters during heating in order to carry out the appropriate treatment of the used oil, to extend the time or change the purpose of use. The basic idea of this research is to improve the quality of edible oils after heating and preparing food. By reducing the content of thermal degradation products extends the usable cycle of the edible oil, with the possibility of reusing it for other purposes.

Changes in the quality of edible oil during heating are considered, based on changes in FFA content, which is one of the basic indicators of oil quality. FFAs present in the oil are the result of hydrolysis - split triglycerides in the presence of water according to the following reaction:

$$\begin{array}{lcl} CH_2OOCR & & CH_2OH \\ \vert & & \vert \\ CH\ OOCR + 3\ H_2O & \longrightarrow & CH\ OH\ +\ 3\ RCOOH \\ \vert & & \vert \\ CH_2OOCR & & CH_2OH \end{array}$$

$$\text{triglyceride} + \text{water} \longrightarrow \text{glycerol} + \text{FFA}$$

Produced FFA in this reaction represents the acidity of the oil that can be expressed as:

- acid number (AN) - mg KOH needed to neutralize FFA in 1 g of oil,
- free fatty acids (FFA) - the mass percentage of FFA in the oil, counts to the "characteristic" fatty acid for concerned type of oil (for sunflower oil acidity is expressed in relation to oleic acid), or
- acid level (AL) - ml NaOH (1 mol/L), necessary to neutralize FFA in 100 g of oil [2].

The FFA content depends on the way of oil production, raw material quality and condition of storage. Freshly prepared refined oils have a smaller FFA content which, during storage and exploitation, increases due to the hydrolysis of oils and producing higher fatty acids as a constituent of higher fatty acid esters [15].

In line with this, the possibility of adsorption of FFA on the natural aluminosilicates was investigated. Adsorption is the one of the most effective and comparable low-cost methods for removing FFA from edible oil [16]. The adsorption process is based on the surface occurrences on the surface of the adsorbent in contact with the gaseous or liquid phase of the adsorbate, whereby the interaction between the adsorbate and adsorbent surfaces is not restricted to the geometric surface. One of the most important requirements for adsorbents is the high surface area of the unit mass or the unit volume of the adsorbent. The surface of the adsorbent is rough, so the reaction also takes place in the interior surface of the walls of the pores, capillaries and cracks.

2 Use of Aluminosilicate in Environmental Protection

After determining the contents of the FFA resulted by heating, the possibility of their adsorption on the natural aluminosilicates, clinoptilolite and bentonite, was researched. The aluminosilicates have been estimated to be suitable for adsorption, because they consist of appropriate system of pores, which size and shape depend on the type or the basic building cell [17].

It is known that a variety of active clays based on aluminosilicate are used as adsorbents during the treatment process of mineral based oils, regeneration of used lubricating oils, bleaching of edible oils, protection of the environment and other. Natural materials used for this purpose are: bentonites, zeolites, bauxites, kaolinite, sepiolite, etc. Their adsorption characteristics depend on the chemical and mineralogical composition as well as on textural, structural and morphological characteristics.

Bentonites are aluminosilicate minerals which, due to their porosity, layered structure and composition, have a wide application. Bentonite is clay of volcanic origin, which as the main component contains clay mineral - montmorillonite with layered structure. It belongs to the subgroup of dioctahedral smectites, whose basic structural unit consists of two tetrahedral coordinated layers of silicon ions, among which one octahedral layer is made of aluminum ions. The composition of montmorillonite except Al, SiO_2 and water is made up of MgO, and sometimes Na or Ca oxide. Besides montmorillonite, bentonite contains other minerals, such as: quartz, calcite, feldspar, muscovite and biotite.

Montmorillonite $Al_2[(OH)_2Si_4O_{10}] \cdot nH_2O$ belongs to a group of layered silicates, whose basic structure is composed of tetrahedral layers, formed by interbinding of tetrahedral sheets with Si^{4+} ion in the center and four O^{2-} ions on the tops, and octahedral layers, formed by binding of octahedral sheets with Al^{3+} ion or Mg^{2+} ion in the center and six OH^- ions on the corners (Fig. 1) [18].

A very important characteristic of bentonite is capability of ion exchange Si^{4+} and Al^{3+} with lower valence cations Na^+, H^+, Ca^{2+}, Mg^{2+}. In addition, bentonite has high adsorption capability, ability of swelling and possibility of processing, which allow its wide range of use. Adsorption capability of natural bentonites can be improved by various methods of activation, modification with various metals, pillaring, etc [19–21]. In Republic of Srpska there are a few deposits of bentonite, and the most important one is located in Sipovo municipality [22].

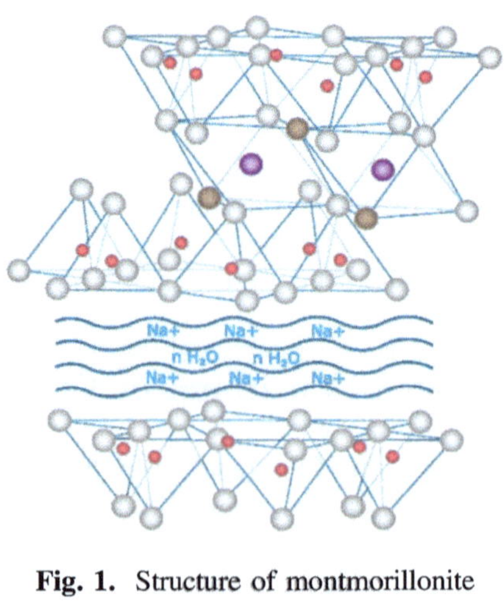

Fig. 1. Structure of montmorillonite

Use of bentonite in environmental protection is wide. They apply to soil protection from erosion, stabilization and recultivation of infertile land and landfills, land protection against pollution, detoxification of drinking water, wastewater treatment, etc.

The naturally occurring zeolite clinoptilolite is one of the most widely used, but also most useful, natural aluminosilicate materials that consist of a three-dimensional framework of SiO_4 and AlO_4 tetrahedra (Fig. 2) in which alkaline and earth alkaline cations are bound. Clinoptilolite structure consists of cavities and channels of molecular level sizes 0.44–0.72 nm and 0.40–0.55 nm. The chemical formula of natural clinoptilolite is $(Na_2\ K_2)\ O \cdot Al_2O_3 \cdot 10SiO_2 \cdot 3H_2O$. It is characteristic that the chemical composition of this mineral usually follows a marked change in Si/Al ratio, although the most common ratio is Si/Al = 4.25/5.25. As a rule, clinoptilolites with low content of Si are enriched with Ca, but often contain Ba and Sr, while clinoptilolites with high content of Si contain Na and K [24].

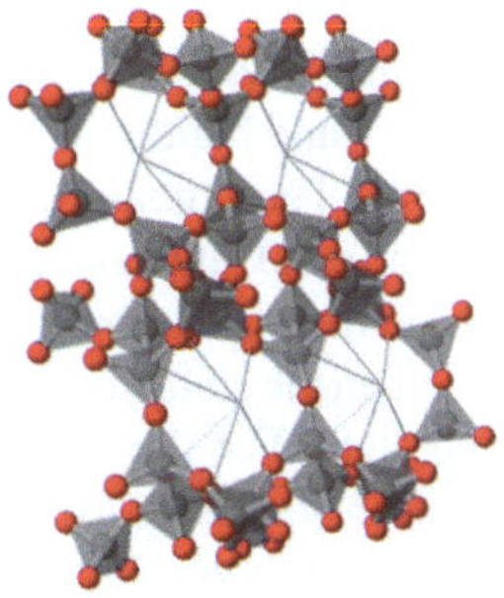

Fig. 2. Structure of clinoptilolite [23]

Nowadays clinoptilolite has found widespread environmental applications which attract the attention of many researchers mainly due to its value to one or more of three properties: adsorption, catalysis and ion-exchange. Major uses of this material include water softening and purification, petrochemical cracking and gas and solvent separation and removal, radioprotection, soil disinfection (after the Chernobyl disaster, the soil around the power plant were treated and neutralized using clinoptilolite), soil improvement, food supplements and additives, biomedical application, etc [25].

Adsorption of the undesirable products of the edible oil heating can contribute to the extension of the possibility of its use. More efficient use or extension of time for the final disposal of used oil as waste, that can no longer be recycled, has importance both economically and ecologically. The enhanced edible oil by adsorption can be used in the manufacture of soaps, lubricants for the food industry, etc [26, 27].

3 Materials and Methods

3.1 Materials

- Sunflower refined edible oil from the market of Bosnia and Herzegovina (one lot of production),
- Bentonite (Sipovo),
- Clinoptilolite - white tuff (Novakovici),
- Mixture diethyl ether: ethanol (95%) = 1:1,
- Standard solution 0.1 mol/L KOH,
- Phenolphthalein indicator,
- $ZnCl_2$.

3.2 Methods

Textural Characteristics of Bentonite and Clinoptilolite
Textural characteristics of clinoptilolite and bentonite were determined by the method of low temperature nitrogen adsorption (LTNA method) on Micromeritics ASAP 2010 at the Laboratory for Physical Chemistry and Catalysis, Faculty of Technology, Novi Sad.

Measurement was perform at liquid nitrogen temperature 77.35 K, in the helium as the carrier gas. The measurement is based on the static monomolecular adsorption of N_2 at the temperature of its liquid aggregate state. For the calculation of the specific surface, a Brunauer-Emmett-Teller-BET equation was used, using a value of $16.2 \cdot 10^{-2}$ nm^2 for a specific surface of an adsorbent molecule, liquid nitrogen [28].

The total pore volume as well as the mean pore diameter were determined based on the adsorption-desorption isotherms (hysteresis) and calculated using the Barrett-Joyner-Halenda (BJH) method [29]. The tested samples were tableted and degassed at 150 °C prior to measurement. Pores were classified according to the Brunauer-Deming-Deming-Teller method based on hysteresis loops of adsorption-desorption isotherms [30, 31].

Determination of FFA Content in Edible Oil
The content of FFA was determined by the method of alkalimetric titration, which is based on the principles of neutralization, and most suitable for oils that are not dyed. In this case, FFA, like weak organic acids, reacts with the alkali and produce soaps. The consumption of the alkali solution is directly proportional to the amount of FFA.

$$CH_3(CH_2)_{16}COOH + NaOH \longrightarrow CH_3(CH_2)_{16}COONa + H_2O$$

stearic acid + alkali ⟶ soap (sodium stearate) + water

After determining the initial reference content of FFA in oil, different samples of oil are heated on a sandy bath at 110 °C, 130 °C, 150 °C, 170 °C and 190 °C during 10 and 30 min. The FFA content was again determined in all samples after heating to determine the influence of temperature. The tests were performed in the Laboratory of the Institute for General and Physical Chemistry, Belgrade.

Adsorption on Bentonite and Clinoptilolite

After heating, the oil was adsorbed on bentonite and clinoptilolite under equal conditions. Adsorption was performed as follows: 50 g of oil sample and 1 g of adsorbents at room temperature (24 °C), stirring at 500 rpm for 60 min. Adsorbents from the adsorbate were separated by centrifugation and filtration. The FFA content was determined again in all samples after adsorption, after which the content of the FFA in filtrate was determined. The tests were performed in the laboratory of the Institute for General and Physical Chemistry, Belgrade.

4 Results and Discussion

Adsorption isotherms of bentonite and clinoptilolite, presented in Figs. 3 and 4, are used to determine the type of adsorption, adsorption capacity, specific surface area and shape of adsorbents pore. Nitrogen on bentonite gives the adsorption isotherm type II, according to IUPAC classification. This isotherm is characteristic for non-porous and macro-porous solids and for substances with mixed micro and meso pores, and it is also corresponds to multimolecular BET adsorption (monolayer coverage followed by multilayering at high relative pressure). Adsorption isotherm on clinoptilolite is type V, convex to the relative pressure axis, characteristic of weak adsorbate-adsorbent interactions at low pressure, microporous and mesoporous solids and hysteresis in multi-molecular adsorption regions.

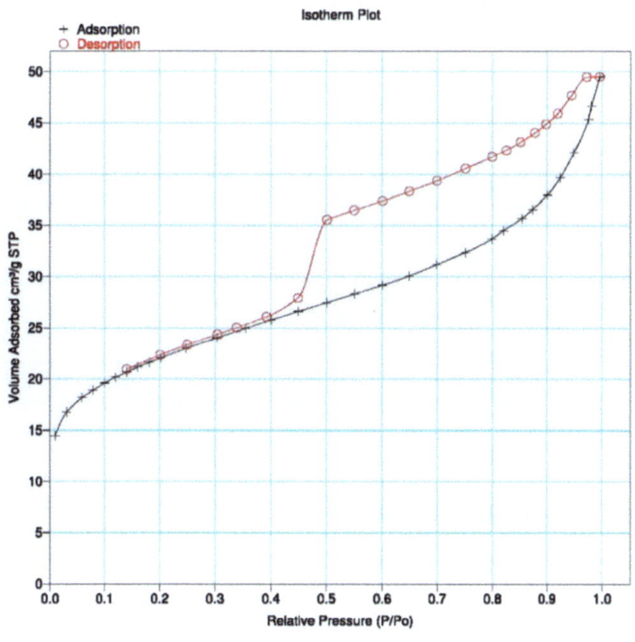

Fig. 3. Adsorption isotherm of nitrogen on natural bentonite at 77.35 K

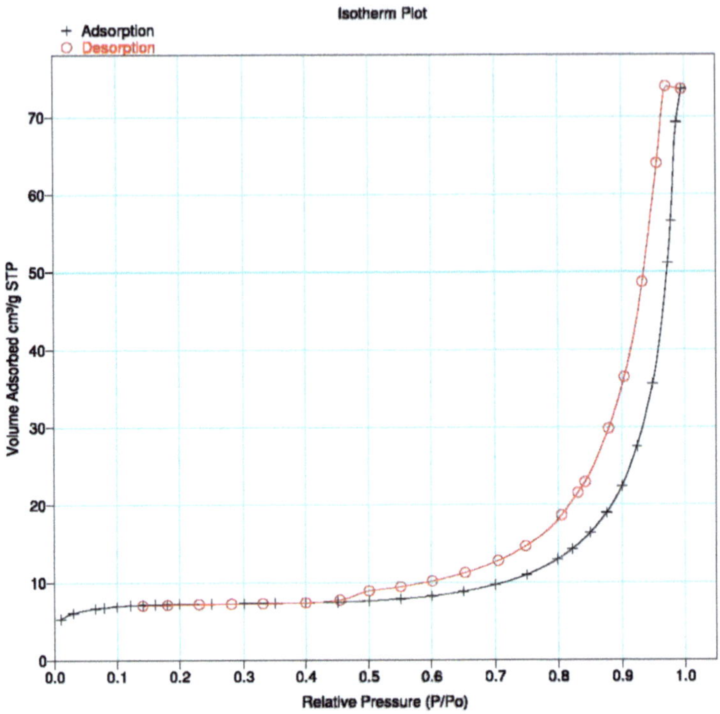

Fig. 4. Adsorption isotherm of nitrogen on clinoptilolite at 77.35 K

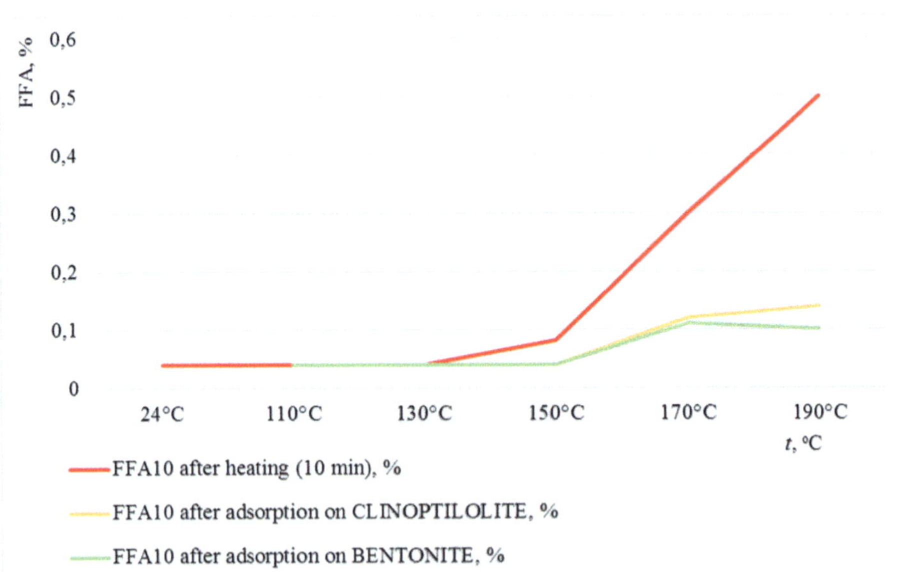

Fig. 5. Effect of ten-minutes heating on FFA content in sunflower oil and recovery of oil after adsorption

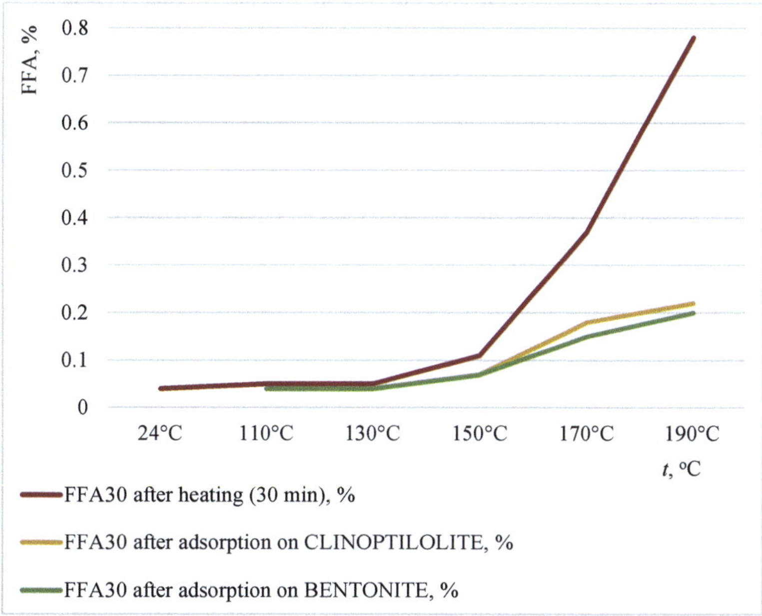

Fig. 6. Effect of 30-min warming on FFA content in sunflower oil and recovery of oil after adsorption

Based on the form of hysteresis loop formed by the adsorption-desorption isotherm of nitrogen, it is possible to assume the shape of the pores in the adsorbent. It can be concluded that bentonite shows H4 hysteresis, according to IUPAC classification. This form is characteristic of complex structures containing both micropores and mesopores and is often associated with narrow slit-like pores. The hysteresis loop in clinoptilolite is H3 type, and their characteristic is that the pores consist of aggregates of plate-like particles which give rise to slit-shaped pores, including cracks [32].

Some important textural characteristics of bentonite and clinoptilolite are given in Table 1. By analyzing these results it can be concluded that the bentonite in relation to clinoptilolite has significantly larger specific BET surface area and external specific surface area, slightly larger microporous volume and microporous surface area, with significantly smaller average pores diameter.

The analysis the FFA content in sample sunflower refined oil before heating shows that FFA content was 0.4%, slightly higher than prescribed by Regulation (max 0.3% FFA, expressed as oleic acid). This is probably caused by long storage and packaging impact.

Experimental results after heating oil samples show that the influence of heating temperature on FFA content in 10-min period is evident at 150 °C (Table 2 and Fig. 5), while at 30-min is evident at 110 °C (Table 3 and Fig. 6). By decomposition of triglycerides, the FFA content increases from 0.04 to 0.5% (after 10 min at 190 °C), or to 0.78% for the continued heating (30 min at 190 °C).

Table 1. Textural characteristics of bentonite and clinoptilolite

Characteristic	Bentonite	Clinoptilolite
Specific BET surface area, m^2/g	77.6005	24.7134
Microporous volume, cm^3/g	0.0105	0.0085
External specific surface area, m^2/g	53.7978	6.1898
Microporous surface area, m^2/g	23.8028	18.5236
Average adsorption pores diameter, nm	5.7306	29.0257
Average desorption pores diameter, nm	4.6102	17.2109

Table 2. Content of FFA in sunflower oil after heating for 10 min and after adsorption on clinoptilolite and bentonite

Heating at temperature	24 °C	110 °C	130 °C	150 °C	170 °C	190 °C
FFA10 after heating (10 min), %	0.04	0.04	0.04	0.08	0.3	0.5
FFA10 after adsorption on clinoptilolite, %		0.04	0.04	0.04	0.12	0.14
FFA10 after adsorption on bentonite, %		0.04	0.04	0.04	0.11	0.10

Table 3. Content of FFA in sunflower oil after heating for 30 min and after adsorption on clinoptilolite and bentonite

Heating at temperature	24 °C	110 °C	130 °C	150 °C	170 °C	190 °C
FFA30 after heating (30 min), %	0.04	0.05	0.05	0.11	0.37	0.78
FFA30 after adsorption on clinoptilolite, %		0.04	0.04	0.07	0.18	0.22
FFA30 after adsorption on bentonite, %		0.04	0.04	0.07	0.15	0.20

Analysis of FFA content after regeneration of the oil by adsorption on bentonite shows that in the oil previously heated at 190 °C for 10 min, the content of FFA decreased by 80% (0.5 to 0.10%) and in the oil heated to 190 °C for 30 min, FFA decreased by 74% (0.78 to 0.20%).

Adsorption on clinoptilolite shows that in the oil previously heated at 190 °C for 10 min, as well as in the oil previously heated for 30 min, FFA content decreased by 72% (0.5 to 0.14% and 0.78 to 0.22%).

5 Conclusion

By analyzing textural characteristics of used adsorbents it can concluded that the bentonite in relation to clinoptilolite has significantly larger specific BET surface area and external specific surface area, slightly larger microporous volume and microporous surface area, with significantly smaller average pores diameter.

The analysis the FFA content in sunflower refined oil before heating shows that FFA content was 0.4%, slightly higher than prescribed by Regulation. The oil samples after heating show that the influence of temperature on FFA content in 10-min period is evident at 150 °C, while at 30-min is evident at 110 °C. By decomposition of triglycerides, the FFA content increases from 0.04 to 0.5% (after 10 min at 190 °C), or to 0.78% for the continued heating (30 min at 190 °C).

Adsorption on bentonite shows that in the oil previously heated at 190 °C for 10 min, the content of FFA decreased by 80%, and in the oil heated to 190 °C for 30 min, FFA decreased by 74%.

Adsorption on clinoptilolite shows that in the oil previously heated at 190 °C for 10 min, as well as in the oil previously heated for 30 min, FFA content decreased by 72%.

Both adsorbents showed good adsorption results. It can be assumed that in bentonite adsorption is mostly carried out on the external surface area, while in clinoptilolite, due to the larger pores diameter, adsorption takes place also on the internal surface. Therefore, approximately equal adsorption results are obtained, although bentonite has bigger specific surface area. For the more accurate estimation of hydrolysis and thermal degradation products it is necessary to analyze the thermally treated oil on the gas chromatograph, as well as thermogravimetric analysis. Further research should be continued on the regeneration of used edible oil by testing other quality parameters, as well as the application of other adsorbents. It is also possible to think about the construction of a pilot plant for the used edible oil regeneration by the adsorption process.

References

1. Orsavova, J., Misurcova, L., Ambrozova, J.V., Vicha, R., Mlcek, J.: Fatty acids composition of vegetable oils and its contribution to dietary energy intake and dependence of cardiovascular mortality on dietary intake of fatty acids. Int. J. Mol. Sci. **16**, 12871–12890 (2015)
2. Dimić, E., Turkulov, J.: Kontrola kvaliteta u tehnologiji jestivih ulja, Tehnološki fakultet Novi Sad (2000)
3. Rossell, J.B.: Frying: Improving Quality. Woodhead Publishing, Cambridge (2001)
4. Martin-Polvillo, M., Marquez-Ruiz, G., Dobarganes, M.C.: Oxidative stability of sunflower oils differing in unsaturation degree during long-term storage at room temperature. J. Am. Oil Chem. Soc. **81**, 577–583 (2004)
5. Vaskova, H., Buckova, M.: Thermal degradation of vegetable oils: spectroscopic measurment and analysis. Proc. Eng. **100**, 630–635 (2015)
6. Fernandes, J.C.B., Draghi, P.F.: Thermal stability of soybean oil: when must we discard it. Food Process. Technol. **2**(5), 00051 (2016)

7. Genic, S., Jacimovic, B., Jaric, M., Budimir, N., Dobrnjac, M.: Research on the shell-side thermal performances of heat exchangers with helical tube coils. Int. J. Heat Mass Transf. **55** (15–16), 4295–4300 (2012). https://doi.org/10.1016/j.ijheatmasstransfer.2012.03.074M20
8. Blagojević, M.S., Blagojević, N.S., Pejić, N., Begović, B., Gajinov, S.: Quality and safety of some commercial spices brands. APTEFF **44**, 1–9 (2013). https://doi.org/10.2298/APT1344001B
9. Sandic, Z., Žunic, M., Maksin, D., Milutinovic-Nikolic, A., Popovic, A., Jovanovic, D., Nastasovic, A.: Glycidyl methacrylate macroporous copolymer grafted with diethylene triamine as sorbent for Reactive Black 5. Hem. Ind. **68**(6), 685–699 (2014). https://doi.org/10.2298/HEMIND140127023SM20
10. Alade, A.O., Jameel, A.T., Muyubi, S.A., Abdul Karim, M.I., Alam, M.D.Z.: Removal of oil and grease as emerging pollutants of concern (EPC) in wastewater stream. IIUM Eng. J. **12** (4), 161–169 (2011)
11. Stams, A.G., Oude, E.S.J.: Understanding and advancing wastewater treatment. Curr. Opin. Biotechnol. **8**, 328–334 (1997)
12. Chao, A.C., Yang, W.: Biological treatment of wool scouring wastewater. J. Water Pollut. Control Fed. **53**(3), 311–317 (1981)
13. Rinzema, A., Boone, M., Knippenberg, K., Lettinga, G.: Bactericidal effect of long chain fatty acids in anaerobic digestion. Water Environ. Res. **66**, 40–49 (1994)
14. Baig, M.A., Mir, M., Bhatti, Z.I., Baig, M.A.: Removal of oil and grease from industrial effluents. EJEAFChe **2**(5), 577–585 (2003)
15. Vasiljević, L.J., Gligorić, M., Dobrnjac, S., Aleksić, V.: Uticaj prženja suncokretovog ulja na sadržaj slobodnih masnih kiselina. V međunarodni kongres, Inženjerstvo, ekologija i materijali u procesnoj industriji, Jahorina (2017)
16. Sandić, Z.P., Nastasović, A.B., Jović-Jovičić, N.P., Milutinović-Nikolić, A.D., Jovanović, D.M.: Sorption of textile dye from aqueous solution by macroporous amino functionalized copolymer. J. Appl. Polym. Sci. **121**, 234–242 (2011). https://doi.org/10.1002/app.33537
17. Kešelj, D.M., Lazić, D.Z., Živković, Ž.D., Škundrić, B.T., Penavin Škundrić, J.V., Sladojević, S.G.: Uticaj stepena kristaličnosti sadržaja aluminijum-oksida i natrijum-oksida na kapacitet sorpcije vode NaY zeolitom. Hem. Ind. **70**(4), 399–407 (2016). https://doi.org/10.2298/HEMIND150126046K
18. Grim, R.E.: Applied Clay Mineralogy. McGraw-Hill Book Company, New York (1962)
19. Zunic, M.J., Milutinovic-Nikolic, A.D., Jovic-Jovicic, N.P., Bankovic, P.T., Mojovic, Z.D., Manojlovic, D.Z., Jovanovic, D.M.: Modified bentonite as adsorbent and catalyst for filtering colour-containing waste waters. Hem. Ind. **645–3**, 193–199 (2012)
20. Kaufhold, S., Dobrmann, R., Klinkerberg, M., Siegesmund, S., Ufer, K.: N_2-BET specific surface area of bentonites. J. Colloid Interface Sci. **349**, 275–282 (2010)
21. Petrovic, R., Dobrnjac, S., Penavin, J., Levi, Z., Bodroža, D., Petkovic, M., Stankovic, T.: Adsorpcija metilen plave boje iz vodene sredine na modifikovanim bentonitima: izoterme i termodinamika. In: Proceedings of XI Conference of Chemists, Technologists and Environmentalists of Republic of Srpska (2016)
22. Mitrovic, D.: Mineralne sirovine Republike Srpske, Republicki zavod za geoloska istrazivanja Republike Srpske Zvornik (2011)
23. http://www.chemtube3d.com/
24. Andrić, L.J.: Fizičko-hemijske karakteristike zeolitskih tufova ležišta Novakovići. Zeoliti Republike Srpske i njihova upotreba u ishrani i poljoprivredi. Univerzitet u Banjoj Luci Rudarski fakultet Prijedor (2011)
25. Bogdanov, B., Georgiev, D., Angelova, K., Yaneva, K.: Natural Zeolites: Clinoptilolite—Review, Economics and Society Development on the Base of Knowledge, Bulgaria (2009)

26. Dobrnjac, S., Vasiljević, L.J., Blagojević, S., Obrenović, Z., Gligorić, M., Tošković, D.: Adsorption of free fatty acids from sunflower oil after heating. In: XX YUCORR Serbia (2018)
27. Wonglamom, J., Rakariyatham, N.: Recovery of used vegetable frying oil by two step adsorbents. In: 26th Annual Meeting of the Thai Society for Biotechnology and International Conference, pp. 330–337 (2014)
28. Skundric, B., Lazic, D., Dobrnjac, S., Petrovic, R., Penavin Skundric, J., Levi, Z., Ostojic, G: Adsorption on Alumina of Different Origin, Contemporary Materials, VII-2, pp. 151–161. https://doi.org/10.7251/comen1602151s (2016)
29. Barret, E.P., Joyner, L.G., Halenda, P.P.: The determination of pore volume and area distributions in porous substances. Computations from nitrogen isotherms. J. Am. Chem. Soc. **73**, 373–380 (1951)
30. Brunauer, S., Emmet, P.H., Teller, E.: Adsorption of gases in multimolecular layers. J. Am. Chem. Soc. **60**, 309 (1938)
31. Lowell, S., Shields, J.E., Thomas, M.A., Thommes, M.: Characterization of porous solids and powders: surface area, pore size and density, vol. XIV. Kluwer Academic Publishers, Dordrecht (2004)
32. Bošković, G.: Heterogena kataliza u teoriji i praksi. Univerzitet u Novom Sadu, Tehnološki fakultet (2007)

Study of Water Vapor Resistance of Co/PES Fabrics Properties During Maintenance

Mirjana Reljic[1,2(✉)], Stanisa Stojiljkovic[3], Jovan Stepanovic[3], Branislava Lazic[2], and Milena Stojiljkovic[3]

[1] CIS Institute, 11000 Belgrade, Serbia
mira.reljic@cisinstitut.rs
[2] The College of Textile - Design, Technology and Management, 11000 Belgrade, Serbia
[3] Faculty of Technology, University of Nis, 16000 Leskovac, Serbia

Abstract. The goal of this study was to investigate the interlinkages of physical, mechanical and thermophysiological properties of fabrics. Changes of a thermophysiological property – water vapor resistance (R_{et}) during washing of six fabrics used for making clothing assemblies (for particular conditions of application) were also examined. Human thermal comfort depends on the combination of clothes, climate conditions and physical activity. Clothing maintenance by washing affects its diffusion and thermal properties. Water vapor permeability and thermal resistance of clothing are key parameter for wearing comfort. Standard methods were applied for examining physical and mechanical properties of fabrics, whereas hot plate measurements were used for testing water vapor resistance (R_{et}). The obtained results indicated that certain yarns properties used in the production of fabrics and constructional characteristics of fabrics significantly affect the water vapor resistance and its changes during washing. The results were used in order to establish a mathematical model for predicting the behavior of the fabrics during washing that are used for a particular clothing assembly in the phase of exploitation – wearing.

Keywords: Thermal resistance · Water vapor resistance · Thermal comfort Maintenance of fabrics

1 Introduction

Thermal comfort depends on the structure of clothing and chemical nature of the fibers (type of fibers and yarns and their properties, fabrics structure and its mechanical and thermal characteristics, finishing, design and clothing fitting. Climate conditions of a clothing wearer are exposed to all human physical activities. These parameters are temperature and relative air humidity, mean radiant temperature and air velocity but also personal factors such as age, gender, metabolism, physical condition, health status etc. [1–5]. Each human physical activity generates certain amount of heat. In order to prevent excessive body heat, i.e. to ensure that body temperature remains constant, excess heat needs to be released [6]. Human body produces moisture in the form of perspiration that needs to leave the microclimate between the skin and fabric before

N. Mitrovic et al. (Eds.): CNNTech 2018, LNNS 54, pp. 72–83, 2019.
https://doi.org/10.1007/978-3-319-99620-2_6

condensation in order to prevent adhesion of the fabric to the skin, thus to keep both the fabric and the skin dry. Velocity and vapor passing through the fabric play an important role in determining thermophysiological comfort, due to affection of human perception. For the thermally comfortable fabric, transmission rate needs to enable higher water vapor. By controlling the properties of vapor transmission, it is possible to develop a wide range of properties of clothing fabrics for different physical activities [7]. Also, it is necessary to design such clothing assemblies which will at the same time protect the wearer from external influences (wind and cold) and provide thermal comfort by carrying water vapor away. Such requirements can be fulfilled by a carefully chosen multi-layered textile structures [8, 9].

Diffusion, adsorption-desorption and convection of vapor perspiration, along with wetting and wicking of liquid perspiration play a significant part in the maintenance of thermophysiological comfort. Diffusion is the main mechanism for transferring moisture in low moisture content conditions. Water vapor diffusion is mainly dependent on the porosity of the fabrics. Convection is important in carrying sweat from the skin to the atmosphere in the case of windy atmospheric condition. With an increase in air velocity, the moisture transfer by convection increases. Wicking plays an important role in moisture transmission, when the moisture content of clothing is very high and the body is producing large quantities of liquid perspiration. Fabric, used for the production of sportswear, should possess very high wicking properties [10].

To achieve good thermophysiological comfort of a clothing wearer, textile materials must possess good properties of moisture absorption, transmission and dissipation. These properties can be influenced by the structure and type of fiber, yarn and fabric, along with finishes and coating applied [11–13].

Cotton fiber is breathable, which means it possesses thermophysiological comfort. There is no static electricity and it is good conductor of heat and electricity. Cotton fabric can absorb high level of moisture. Unfortunately, the wicking property between inner and outer surfaces of fabrics made of cotton fibers is very weak, which makes cotton inadequate to wear next to the skin during energy-burning activities. Polyester fiber has a very low absorption property. When one perspires, polyester tends to keep the perspiration trapped against the body. Due to the hydrophobic nature, polyester is also more electrostatic compared to the natural fibers [11]. Since moisture management depends on the materials and their blend in the fabric, combinations of hydrophilic and hydrophobic materials were selected in order to improve moisture management, such as Co/PES blend [11, 14], or a textile material is being treated [13].

Besides achieving required thermal comfort, it is particularly important to extend the life span of clothes as much as possible, in order to produce clothing assemblies whose thermal comfort, i.e. diffusive and thermal properties will not significantly change during usage. Water vapor permeability and thermal comfort of clothing are key parameters for the assessment of clothing in the exploitation/wearing phase [2]. For effective use of clothing, it is important to predict the period of maintenance of functional materials properties it is made of. After this period, certain functional properties such as thermal comfort will be lost due to maintenance/washing.

Since water vapor permeability is a very important parameter of thermal comfort of clothing, the paper investigated the change of water vapor resistance (R_{et}) during washing of six fabrics used for the production of clothing assemblies for a particular wearing situation. The aim of the research was to establish functional dependence of R_{et} values on the number of washing cycles, which would enable prediction of the behavior of a clothing assembly during its usage (determination of its life span for a particular purpose).

2 Experimental Part

2.1 Materials

The tests were carried out on a group of cotton fibers BOX 4, group of polyester fibers ESLON, four articles yarn in the mixture 50% Co −50% PES and six fabrics 50% Co −50% PES samples of fabrics produced by Holding company "Yumco" (Vranje, Serbia), but no typical ones were selected for the research – six fabrics (Table 1). The selected fabrics were made of the blend of cotton (BOX 4) fibers of medium length and polyester fibers (ESLON).

Table 1. Labels of fabric samples

Samples	Product name	Composition Co/PES, %
1	R-37-46842	51.6/48.4
2	6B15046848	50.2/49.8
3	6B15046920	50.7/49.3
4	R-37-46842 row	51.8/48.2
5	6B15046737	50.0/50.0
6	6B15046961	50.2/49.8

ESLON fiber is colored without using any special carriers, at the 305 bar and 130 °C, with dispersed dye in HT apparatus (High Speed Machine for dyeing, ZMY, China), which represent one of the advantages in the application of this fiber. ESLON fiber is also very similar to the cotton fiber BOX 4 (vertical section), thus one more benefit for the production of blend of these two fibers is achieved.

3 Methods

Parameters of Co and PES fibers (BOX 4 and ESLON, respectively), Co/PES yarns used for the production of clothing fabrics and properties of fabrics prior to first washing cycle were tested. Also, change of mass per unit area of fabrics at washing (up to 50 washing cycles) and change of the properties of water vapor permeability and thermal resistance of fabrics expressed through R_{et} values during maintenance, i.e. washing of fabrics (up to 50 washing cycles) were examined.

The experiment was carried out on the hotplate (ELEKTRO UMI, Serbia) and in the washing machine (Elektrolux, WASCATOR FOM 71 CLS). Washing procedure of fabric 4N-Normal was applied (Table 2), according to ISO6330: 2015 (E) standard. The experimental part was carried out in CIS Institute (Belgrade, Serbia).

Table 2. Washing procedures for reference washing machine Type A

Procedure No.	Agitation during heating, washing and rinsing	Washing				Rinse 1		Rinse 2		Rinse 3			Rinse 4	
		Temp.	Liquor level	Wash time	Cool down	Liquor level	Rinse time	Liquor level	Rinse time	Liquor level	Rinse time	Liquor level	Rinse time	Spin time
		a	bc	d	f	bc	dg	bc	dg	bc	dg	bc	dg	d
		°C	mm	min		mm	min	mm	min	mm	min	mm	min	min
4 N	Normal	40 ± 3	100	15	No	130	3	130	3	130	2	130	2	5

Standardized methods were used in this research for characterizing and examining the quality of fibers, yarns, fabrics. Methods of biophysical analysis of clothing on a hot plate were used. Also, adequate mathematical methods for processing experimental results were applied.

Changes of the mass per unit area of the fabrics during 50 washing cycles were measured according to ISO 3801:1977 standard.

3.1 Hot Plate Method

Measurements of water vapor resistance (R_{et}) values of the fabrics were carried out by means of hot plate method (sweating guarded hotplate test). The advantages of this method are easy execution and handling, quick individual measurement and good repeatability of the measurement. A hot plate simulates processes which occur along human skin [15].

For the determination of water vapor resistance, an electrically heated porous plate is covered by a water vapor permeable but liquid water impermeable membrane. Water feeds the heated plate with the vapor which passes through the membrane so that no liquid water contacts the test sample. When the test sample is placed on the membrane, the heat flux required to maintain the hotplate temperature is a measure of the rate of water evaporation. It is determined properly from water vapor resistance of the sample.

During the water vapor resistance test, the power required to maintain the plate temperature at 35 °C, relative air humidity at 40%, and air flow velocity $v = 1.0\ \text{ms}^{-1}$ is being measured. Water is supplied to the surface of a porous plate 1 by a dosing device 5, such as a motor-driven burette. In order to maintain a constant rate of evaporation, the dosing device is activated by a switch. These switch senses when the level of water in the plate falls more than 1 mm below the plate surface. The level switch is mechanically connected to the measuring unit. Before entering the measuring unit, water is preheated to the temperature of the measuring unit. This can be achieved by passing it through the tubes in thermal guard before it enters the measuring unit. Water vapor resistance of the fabric is expressed in unit m^2Pa/W.

This method enables water vapor resistance (R_{et}) of a material to be determined by subtracting the water vapor resistance of the boundary air layer.

3.2 Comfort Rating According to R_{et} Values

Classification of fabrics was carried out according to the system by Hohenstein Institute, which classify clothing comfort in accordance with R_{et} values (Table 3) [16]. Thermal comfort of the clothing during exploitation can be predicted based on the measured water vapor resistance.

Table 3. Ret comfort rating system [6]

Rating	R_{et} value	Description
Very good	0–6	Extremely breathable and comfortable at a higher level of activity
Good	7–13	Very breathable and comfortable at a moderate rate of activity
Satisfactory	14–20	Breathable, but uncomfortable at a higher rate of activity
Unsatisfactory	21–30	Slight breathable, giving moderate comfort at low rate of activity
Very unsatisfactory	31+	Not breathable and uncomfortable, with a short tolerance time

3.3 Establishing the Mathematical Model of Dependence of R_{et} Values on the Number of Washing Cycles

For determining dependence of the tested textile R_{et} values of the materials on the number of washing cycles, an adequate mathematical model was proposed. In order to illustrate a non-linear regression model, a polynomial model was presented as it was much simpler than the modified gradational model [17, 18].

For the given string of data pairs (x_1, y_1), (x_2, y_2), …, (x_n, y_n), dependence of the form $y = a_0 + a_1x + \ldots + a_mx^m$ needs to be determined, where $m \leq n - 2$, so as to illustrate the experimental data in the best way possible.

As a quality measure of fitting the experimental data obtained by the empirical formula, mean square deviation of the formula from the experimental values is used, and it is defined as:

$$S = \frac{\sum_{i=1}^{n} E_i^2}{n-(m+1)} = \frac{\sum_{i=1}^{n} (y_i - f(x_i))^2}{n-(m+1)} \tag{1}$$

Where:

- $f(x_i)$: – value of approximated function in every point,
- $n - (m + 1)$ – value representing the number of freedom degrees defined as a difference between the number of experimental points and the total number of parameters.

The smaller the value S, the better the empirical formula fits experimental data.

4 Results and Discussion

Results for fiber parameters are given in Table 4, results for the Co/PES yarns in Table 5, while Table 6 provided results for the fabric parameters.

Table 4. Comparison of length, fineness and breaking force of cotton and polyester fiber (BOX 4, ESLON)

Property	Test method	Unit	Co BOX 4	PES ESLON
Composition	–	%	100	100
Length	–	mm	27/28	38
Fineness	SRPS EN ISO 2060:2012	mikroner dtex	4.6 1.81	1.4
Breaking force	SRPS EN ISO 2062:2012	psi cN/tex	8.5 P.I. 92000 44.71	56.60

Table 4 provides a comparative survey of physical and mechanical properties of cotton and polyester fibers (length, fineness and breaking forces). It can be noticed that the values of selected fibers BOX 4 and ESLON are close and very compatible. The best linearity of the fabric is achieved in this way, so they are most often used in the blend.

Mass per unit area of tested fabrics before washing and changes in surface mass of fabrics at a particular number of washing cycles is presented in Table 7.

R_{et} values of the fabrics before washing (Table 8) range between 1.76 m^2Pa/W and 4.33 m^2Pa/W and are shown in, as well as the changes of R_{et} values at a certain number of washing cycles.

According to the classification by the Hohenstein Institute, all tested fabrics show a small degree of water vapor resistance before washing (R_{et} values range between 1 and 6 m^2Pa/W – very good or extremely comfortable during high-intensity physical activities) [6]. The highest R_{et} value, i.e. the lowest water vapor permeability and consequently the lowest thermal comfort is observed in sample 5, whereas the smallest R_{et} value, i.e. the highest thermal comfort is observed in sample 2. This is expected since sample 5 is made of the finest yarn and has the highest density of threads and mass per unit area, whereas the smallest mass per unit area is observed in sample 2 (Table 6).

Based on experimental results, it can be concluded that mass per unit area has a strong impact on water vapor resistance – the thinner the mass per unit area, the smaller the R_{et} value is.

With washing, a slight decrease of mass per unit area can be observed in all samples, even after 50 washing cycles. Water vapor resistance increase due to samples washing and their relative relation did not change.

So, by observing the changes in properties of the tested fabrics at 50 washing cycles, it is found that a change occurred in the thermophysiological R_{et} property. Water vapor resistance value, R_{et}, increased with the number of washing cycles in all samples (Table 8).

After 50 washing cycles, changes in water vapor resistance occurred due to shrinking of the fabric, i.e. a slight increase in the density of threads. However, the measured deviations are not large and they do not have a significant impact on the thermophysiological properties of clothing.

Table 5. Breakdown of properties Co/PES yarns

Property	Test method	Unit	Article (type of medium length cotton fiber Box 4)			
			Art. 650/50	Art. 600/40	Art. 14/24	Art. 16/34
Composition	ISO 1833-1:2006 ISO 1833-11:2006 ISO 1833-24:2009	%	50% Co 50% PES	50% Co 50% PES	50% Co 50% PES	50% Co 50% PES
Breaking force	SRPS EN ISO 2062:2012	cN	368.8	490.8	348.0	362.0
Breaking elongation	SRPS EN ISO 2062:2012	%	7.8	7.1	6.4	5.5
Direction of twist and number of twists per unit of length in a yarn	SRPS EN ISO 2061:2012	u/m	1x-756Z 2x-192S	1x-674Z 2x-171S	1x-665Z	1x-660Z
Fineness	SRPS EN ISO 2060:2012	tex	20.0x2	25.0x2	41.7x1	29.4x1
Smoothness on Uster	ISO 16549:2004					
- Thin places		number	5	/	/	1
- Thick places		number	85	28	35	48
- Number of lumps		number	45	33	48	31
- U (Uster)		%	14.6	12.1	10.1	10.7

Table 6. Breakdown of properties of fabrics made of Co/PES mixture

Property/ test method	Unit	Fabric sample					
		1	2	3	4 row	5	
Warp yarn, Art.		16/34	16/34	600/40	16/34	650/50	650/50
Weft yarn, Art.		16/34	16/34	14/24	16/34	14/24	650/50
Composition -cotton ISO 1833-11:2006	%	51.6	50.2	50.7	51.8	50.0	50.2
-polyester ISO 1833-24:2010		48.4	49.8	49.3	48.2	50.0	49.8
Thickness ISO 5084:1996	mm	0.67	0.42	0.60	0.43	0.58	0.45
Mass per unit area ISO 3801:1977	g/m^2	252.7	208.6	298.1	236.8	335.7	266.8
Number of threads per unit length -warp -weft ISO 7211-2:1984	cm^{-1}	48.0 26.0	44.0 21.0	30.9 26.0	44.3 25.6	50.0 27.0	34.7 28.7
Yarn fineness -warp -weft SRPS ISO 2060:2012	tex	30.7x1 29.8x1	28.9x1 29.6x1	25.8x1 41.7x1	29.6x1 29.5x1	20.0x2 41.4x1	20.6x2 21.8x2
Weave structure – twill		3:1	2:1	3:1	3:1	3:1	2:2

Table 7. Changes of mass per unit area of fabrics at a different number of washing cycles (measured according to method ISO 3801:1977)

Sample	Surface mass of fabric (g/m^{-2})										
	Untreated	After 5 washings	After 10 washings	After 15 washings	After 20 washings	After 25 washings	After 30 Washings	After 35 washings	After 40 washings	After 45 washings	After 50 washings
1	252.7	252.6	252.5	252.1	252.0	251.9	251.7	251.6	251.5	251.3	251.3
2	208.6	208.4	208.0	207.3	207.1	206.9	206.6	205.9	205.8	205.4	205.3
3	298.1	297.9	297.9	297.5	297.3	297.1	296.9	296.7	296.7	296.6	296.5
4	236.8	236.8	236.7	236.6	236.6	236.2	235.9	235.8	234.5	233.7	233.2
5	335.7	335.6	335.6	335.3	335.1	334.9	334.8	334.7	334.6	334.2	334.0
6	266.8	266.7	266.6	266.4	266.3	266.1	265.7	265.5	265.3	265.2	264.9

Table 8. Changes of R_{et} values at 50 washing cycles of fabric samples 1–6

Sample	R_{et} (m^2 Pa/W)										
	Untreated	After 5 washings	After 10 washings	After 15 washings	After 20 washings	After 25 washings	After 30 washings	After 35 washings	After 40 washings	After 45 washings	After 50 washings
1	3.35	3.75	4.72	4.91	5.25	5.91	6.43	6.75	7.15	7.35	7.25
2	1.76	1.92	2.60	3.32	3.25	3.63	3.85	4.10	3.97	4.05	4.12
3	3.79	3.85	3.87	4.32	4.15	4.53	4.45	4.97	5.23	5.73	5.63
4	2.90	2.90	3.00	4.10	4.70	4.60	4.90	5.00	5.00	5.00	5.20
5	4.33	4.95	6.15	6.25	6.25	6.15	6.75	6.99	7.53	7.71	7.89
6	2.98	2.99	3.51	4.05	4.15	4.63	4.90	4.97	5.10	5.25	5.41

McCullough and Jones defined resistance to the heat-carrying of dry clothes [19, 20]. Fanger used a simple dry insulation model [21]. Kerslake gave a model where the latent heat of evaporation is taken out of clothing, but not from the skin [22]. Changes in the water vapor resistance in the washing process have not been studied so far. The thermophysiological comfort of textile materials was not sufficiently investigated. Therefore, thermophysiological comfort of textile materials with one-time wear, as well as after use, or adequate maintenance of the finished product [23–25] was main contribution of this study.

4.1 Functional Dependence of R_{et} Values on the Number of Washing Cycles

The results obtained by examining the changes of R_{et} values in respect to the number of washing cycles are approximated by a non-linear model of data fitting in the form:

$$R_{et} = a + b \cdot n^{c} \tag{2}$$

where a, b and c are constants that are determined to obtain local optimal matching with the given data (this property is pointed out since linear models require global matching with the data), whereas n is an independent variable and present the number of washing cycles a sample is subjected to.

Modified gradational model was used for the approximation purposes due to experimental results that follow the gradational function and their "untreated" initial value. Modification is reflected by adding the constant a.

Table 9 provides a breakdown of calculated coefficients of the empirical formula and a root mean square deviation (S) value for the changes of R_{et} values in respect to the number of washing cycles in the tested samples. Since S values are very small for all samples, it can be concluded that empirical formula that describe functional dependence of R_{et} values on the number of washing cycles, properly fit experimental data for all samples. Also, the suggested mathematical model can be applied to the similar fabrics.

Table 9. Breakdown of calculated coefficients of the empirical formula and breakdown of S values for changes of Ret values depending on the number of washing cycles of fabric samples 1–6

Sample	$R_{et} = a + b \cdot n^c$			S
	a	b	c	
1	2.87080	0.37990	0.64350	0.64350
2	0.67411	0.92183	0.34995	0.34995
3	3.79190	0.00427	1.57170	1.57170
4	2.15970	0.48147	0.48563	0.48563
5	3.78110	0.61448	0.47849	0.47849
6	2.55600	0.25657	0.62476	0.62476

The summative overview of all data obtained by measurements and a polynomial by means they were approximated for different values of a and b constants and exponent c, are presented in Fig. 1.

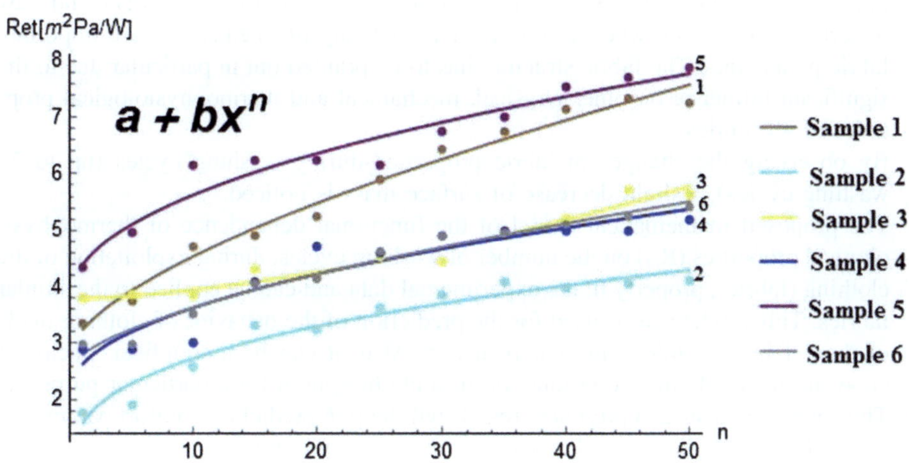

Fig. 1. Functional dependence of ret values on the number of washing cycles of fabric samples

From S value on the samples 1–6, it can be concluded that the empirical formula best fit the experimental data on sample number 2 due to increased R_{et} value by 234.09% after 50 washing cycles. This value represents the highest change of R_{et} value in maintenance for all samples. After 50 washing cycles, R_{et} value of sample number 3 showed lowest increase of all samples by 148.54%. There was a significant deviation in R_{et} value after 10 washings in sample 5 and sample 6. Sample 1 and sample 3 have identical weave structure and approximately the same value of fabric thickness. Also, mass per unit area decreased by about the same value during the maintenance process for both samples.

Figure 1 shows that the fabric sample number 2 has the lowest R_{et} value (4.12 m^2Pa/W) after 50 washing cycles. It means that sample number 2, compared to other tested samples, has the highest resistance to water vapor flow. Therefore, the most pleasant fabric for carrying after 50 washing cycles would be sample number 2.

5 Conclusion

Based on the analysis, comparison and statistical processing of the obtained results relating to the thermos-physiological property of materials – water vapor resistance, the following can be concluded:

- For the quality depiction of the properties of textile products water vapor resistance, a systematic approach which includes measurement and calculation of a number of parameters relating to fiber-yarn-fabric-finishing-clothing sequence needs to be applied.
- The conducted tests indicate that certain properties of yarn and fabrics significantly affect the change in water vapor resistance of the fabrics. Of the yarn parameters, important are the fineness (longitudinal mass), length of yarn fibers and raw material composition, whereas surface mass, density of threads, etc. are important fabric parameters. The fabric structure has to be pointed out in particular due to the significant influence on other physical, mechanical and thermophysiological properties of all fabrics.
- By observing the changes of fabric properties during washing cycles (up to 50 washing cycles), a slight decrease of surface mass is noticed.
- The proposed mathematical model of the functional dependence of thermophysiological properties (R_{et}) on the number of washing cycles, during exploitation of the clothing (fabric), properly fit the experimental data and can be applied to the similar fabrics. This model can be used for the prediction of the behavior of clothing made of these fabrics, while being cared in use. Also, it can be a significant factor in choosing materials for the production of a clothing item for a particular purpose.
- The conducted tests and presented results will help in predicting comfort parameters of similar clothing items and assessing their usage properties.

References

1. Long, H.: Int. J. Cloth. Sci. Technol. **11**, 198–205 (1999)
2. Ruckman, J.E., Murray, R., Choi, H.S.: Int. J. Cloth. Sci. Technol. **11**, 37–52 (1999)
3. Tuğrul Oğulata, R.: Fibres Text. East. Eur. **15**(2), 67–72 (2007)
4. Humidity and Its Impact on Human Comfort and Wellbeing in Occupied Buildings, Humidity Group of the HEVAC Association (2016)
5. https://www.researchgate.net/publication/270837142. Access 04 Feb 2017
6. Özkan, E.T., Meriç, B.: Text. Res. J. **85**(1), 62 (2015)
7. Ramkumar, S.S., Purushothaman, A., Hake, K.D., McAlister, D.D.: J. Eng. Fibers Fabr. **2** (4), 10 (2007)
8. Farzandi, M., Razipour, S., Mousazadegan, F., Saharkhiz, S.: Int. J. Cloth. Sci. Technol. **25** (2), 99 (2013)
9. Bivainyte, A., Mikučioniene, D., Kerpauskas, P.: Investigation and thermal properties of double-layered weft knitted fabrics. Mater. Sci. (Medžiagotyra) **18**(2), 167–171 (2012)
10. Das, B., Das, A., Kothari, V.K., Fanguiero, R., de Araújo, M.: AUTEX Res. J. **7**(2), 100 (2007)
11. El Messiry, M., El Ouffy, A., Issa, M.: Alex. Eng. J. **54**, 127 (2015)
12. Tyagi, G.K., Krishna, G., Bhattacharya, S., Kumar, P.: Indian J. Fibre Text. Res. **34**, 137–143 (2009)
13. Rosace, G., Guido, E., Colleoni, C., Barigozzi, G.: Int. J. Polym. Sci. **2016**, 10p. (2016). Article ID. 1726475
14. Pant, S., Jain, R.: Stud. Home Comm. Sci. **8**(2–3), 69–74 (2014)
15. Salmon, D.: Meas. Sci. Technol. **12**, R89–R98 (2001)
16. Williams, J.T.: Textiles for Cold Weather Apparel, p. 159. Woodhead Publishing in Textiles (2009)
17. Herceg, D., Krejić, N.: Numerical Analysis. Stylos, Novi Sad (1997)
18. Milovanović, G.V.: "Numerical analysis" (I, II and III part). Naučna knjiga, Belgrade (1985)
19. McCullough, E.A., Jones, B.W.: A comprehensive database for estimating clothing insulation. IER Technical report 84-01, Institute for Environmental Research, Kansas State University, December 1984
20. ISO 9920:1995 Ergonomics of the thermal environment - Estimation of the thermal insulation and evaporative resistance of a clothing ensemble
21. Fanger, P.O.: Thermal Comfort. Danish Technical Press, Copenhagen (1970)
22. Kerslake, D.M.: The Stress of Hot Environment. Cambridge University Press, Cambridge (1972)
23. Popov, D.B., Ćirković, N., Stepanović, J., Reljić, M.: The analysis of the parameters that influence the seam strength. Ind. Text. (3), 131–136 (2012)
24. Stepanovic, J., Ćirkovic, N., Radivojevic, D., Reljic, M.: Defining the warp length required for weaving process. Ind. Text. **63**(5), 227–231 (2012)
25. Asanović, K.A., Cerović, D.D., Mihajlović, T.V., Kostić, M.M., Reljić, M.M.: Quality of clothing fabrics in terms of their comfort properties. Indian J. Fibre Text. Res. **40**(4), 363–372 (2015)

Experimental Dimensional Accuracy Analysis of Reformer Prototype Model Produced by FDM and SLA 3D Printing Technology

Aleksa Milovanović[1(✉)], Miloš Milošević[1], Goran Mladenović[2], Blaž Likozar[3], Katarina Čolić[1], and Nenad Mitrović[2]

[1] Innovation Centre of the Faculty of Mechanical Engineering, University of Belgrade, 11120 Belgrade, Serbia
amilovanovic@mas.bg.ac.rs

[2] Faculty of Mechanical Engineering, University of Belgrade, 11120 Belgrade, Serbia

[3] Department of Catalysis and Chemical Reaction Engineering, National Institute of Chemistry, 1001 Ljubljana, Slovenia

Abstract. The subject of this paper is the evaluation of the dimensional accuracy of FDM and SLA 3D printing technologies in comparison with developed reformer polymer electrolyte membrane (PEM) fuel cell CAD model. 3D printing technologies allow a bottom-up approach to manufacturing, by depositing material in layers to final shape. Dimensional inaccuracy is still a problem in 3D printing technologies due to material shrinking and residual stress. Materials used in this research are PLA (Polylactic Acid) for FDM technology and the standard white resin material for SLA technology. Both materials are commonly used for 3D printing. PLA material is printed in three different height resolutions: 0.3 mm, 0.2 mm and 0.1 mm. White resin is printed in 0.1 mm height resolution. The aim of this paper is to show how layer height affects the dimensional accuracy of FDM models and to compare the dimensional accuracy of FDM and SLA printed reformer models with the same height resolution.

Keywords: Reformer · 3D printing · FDM · SLA · CAD model

1 Introduction

3D printing technology allows fabrication of objects through a sequential layering of material based on digital CAD model, [1, 2]. A variety of 3D printing processes are now commercially available, differing from each other in the way they build a model. In the process of 3D printing, the material is softened, melted or irradiated in order to deposit material layer by layer, [3]. As a technology which can achieve production of complex geometry parts, first 3D printing was developed for the purpose of rapid prototyping. Nowadays, there are examples of 3D printing technologies used for specialized customer products, structural models in architecture, aerospace and medical industry with the tendency to wider use in industrial applications, [2, 3].

N. Mitrovic et al. (Eds.): CNNTech 2018, LNNS 54, pp. 84–95, 2019.
https://doi.org/10.1007/978-3-319-99620-2_7

Geometrical inaccuracy from shrinkage and residual stress induced deformations are key sources of defects in 3D printing, [4]. Material shrinkage has been a major cause of part inaccuracy. Most common and commercially available 3D printing technologies are FDM (fused deposition modeling) and SLA (stereolithography), [2]. In this paper, FDM and SLA technology would be compared in terms of dimensional accuracy of finished parts based on digital CAD models of PEM fuel cell reformer, [5]. 3D printing is still in early stages of commercialization and thus presents a high potential for research and development, [6].

FDM is a material extrusion 3D printing technology in which softened thermoplastic material is extruded through a nozzle or orifice onto a platform (Fig. 1). FDM machine is equipped with computationally controlled extruder mechanism. The platform can be equipped with resistant heaters in order to keep extruded material at certain temperature holding material on the platform. Printing material in form of a filament is guided by two pulleys into the extruder. At the extruder end heater block heats and keeps material at a temperature above melting point allowing a steady flow of material filament through the nozzle. Nozzle extrudes softened material on a pre-defined path, creating the first layer. Thereafter, the platform moves downwards allowing the nozzle to generate next layer. Printing continues, layer-by-layer, until a model is finished, [1, 2, 7]. Dimensional accuracy of a printed model is determined by layer height and nozzle diameter more than extrusion temperature, printing speed and infill percentage and pattern, [8].

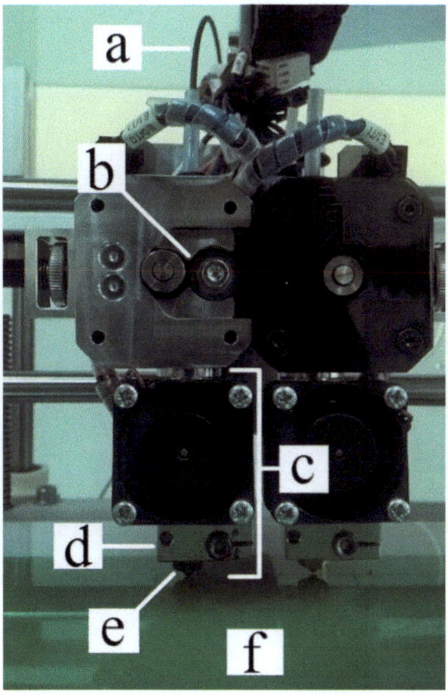

Fig. 1. FDM printing technology: (a) material filament; (b) two pulleys; (c) extruder; (d) heater block; (e) nozzle; (f) platform/heated bed.

FDM uses a vast variety of thermoplastic materials with different mechanical properties, which makes this particular technology favorable for industrial use. Most common materials in FDM technology are ABS (Acrylonitrile Butadiene Styrene) and PLA (Polylactic Acid) polymer. PLA is environmental friendly biodegradable plastic in comparison with petroleum-based ABS, [9]. The melting point of PLA is at 175 °C, [10].

During printing, heated material cools at a lower temperature of platform/heated bed causing changes in material volume, i.e. shrinking, which leads to dimensional inaccuracies as opposed to the digital CAD model, [11, 12]. Model accuracy depends on the physical properties of the material during heating and cooling, [3]. PLA has proven to be more dimensionally stable after print compared to ABS material, which is why this material is a subject of this study, [9, 10].

SLA (Stereolithography) is vat photopolymerization 3D printing technology in which liquid photopolymer is selectively irradiated by a laser beam causing light-activated polymerization of liquid material (Fig. 2), [1, 3]. Main components of SLA machines are a light source, build platform and resin tank. Liquid photopolymer-resin is stored in resin tank with a transparent bottom. UV laser is used as a light source to initiate the polymerization of liquid material. Print starts by lowering of the build platform into the resin tank, leaving space between bottom of the tank and build platform equal to layer height. UV laser directs the light through the bottom of the resin tank selectively curing a layer of material. When laser finishes irradiating the first layer, the platform moves up to allow wiper to pass across the tank to circulate resin. The process continues layer-by-layer until the model is printed, [1–3]. The whole process is enclosed to prevent other light sources from curing the resin material.

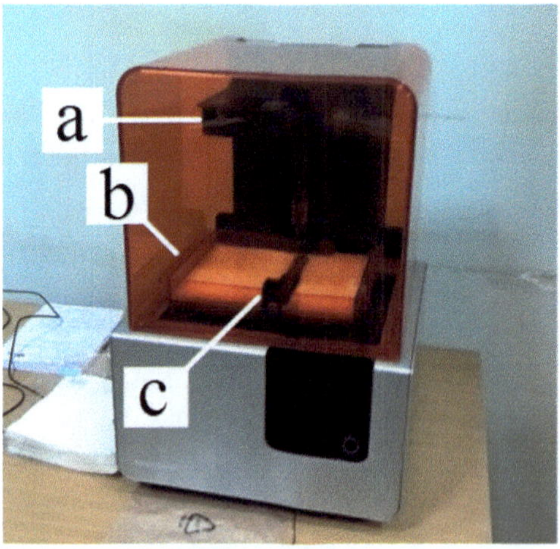

Fig. 2. SLA printing technology: (a) build platform; (b) resin tank; (c) wiper.

The resin material used in this research is white standard resin developed by Formlabs Inc. (Somerville, MA). The resin material is thermoset plastic which remains in solid state after polymerization, [13]. SLA prints have a smooth finish, with no visible layer lines, [2]. SLA prints have dimensional inaccuracies due to heating and cooling of the cured resin during the printing process, [3, 14].

2 Materials and Methods

In this research, PEM fuel cell reformer model is used as a geometric benchmark for comparing the dimensional accuracy of FDM and SLA 3D printing technologies. Reformer model can benchmark flat surfaces, circular features (holes and cylinders) and surface finish.

Benchmarking is used to compare different similar systems to establish a standard of performance. There are three types of benchmarking in 3D printing: geometric, mechanical and process benchmarking. Geometric benchmarking is used to compare dimensional accuracy, surface finish, flatness and straightness of printed models, [6]. Geometric benchmarking in 3D printing has a wide application in present research. Recent research of Sljivic et al. [4] compared the dimensional accuracy of consumer grade and professional FDM 3D printer with complex three-dimensional model of one particular cathedral located in Bosnia and Hercegovina, [4]. Queral et al. examined dimensional accuracy of several 3D printing technologies (i.e. FDM, SLA, SLS-Selective Laser Sintering, PolyJet) for fabrication of modular coil frames on stellarator device, [15]. Ogden et al. analyzed the influence of process parameters on dimensional accuracy of 3D printed human vertebra which can improve 3D printing for medical and tissue engineering applications, [16]. One of the aims of benchmarking is to find best possible technology and solution for fabrication of non-conventional parts that are not practical to be produced by standard machining methods. One of such models is the subject of this paper, i.e. polymer electrolyte membrane (PEM) fuel cell reformer.

Reformer is the first main reactor of a PEM fuel cell. Reformer consists of three plates-upper and lower cover plate and middle plate. Reaction volume of the reformer is placed in the middle plate, with inlet and outlet channel-both with a diameter of 3 mm. Reaction volume dimensions are $34 \times 37.4 \times 4$ mm. The thickness of lower, middle and upper plates are 3.8 mm, 4 mm and 1 mm, respectively. Lower plate contains the hole, 6 mm diameter, for filling the catalyst. Plates have two channels, 5 mm diameter, for cartridges that connect all three plates. External dimensions of all plates are 60×60 mm, [17]. Digital CAD models of all components of PEM fuel cell reformer are developed in SolidWorks software (shown on Figs. 3, 4, 5 and 6).

FDM printer German RepRap X400 (Feldkirchen, Germany) and SLA printer Formlabs Form2 (Somerville, MA) were used for this research in order to make PEM fuel cell reformer prototypes. Digital CAD models of reformer were converted to STL format, which uses triangular facets to approximate the shape of an object, in order to prepare models for adequate slicing software. For FDM printer Simplify3D® slicer software (Cincinnati, OH) is used for generation of G-code. SLA printer uses PreForm (Formlabs, Somerville, MA) software for preparation of printing process.

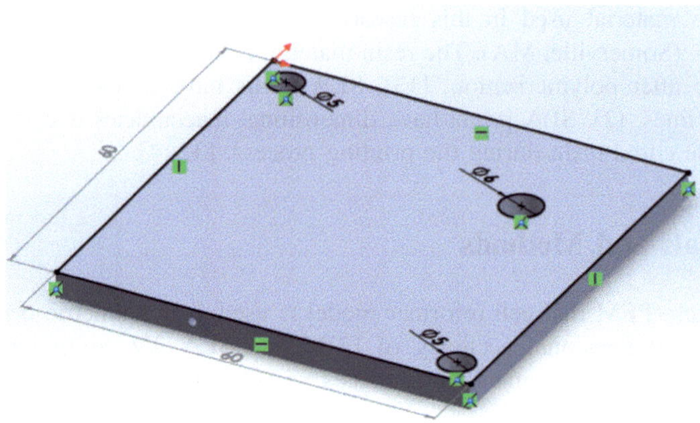

Fig. 3. Digital CAD model of reformer PEM fuel cell lower plate.

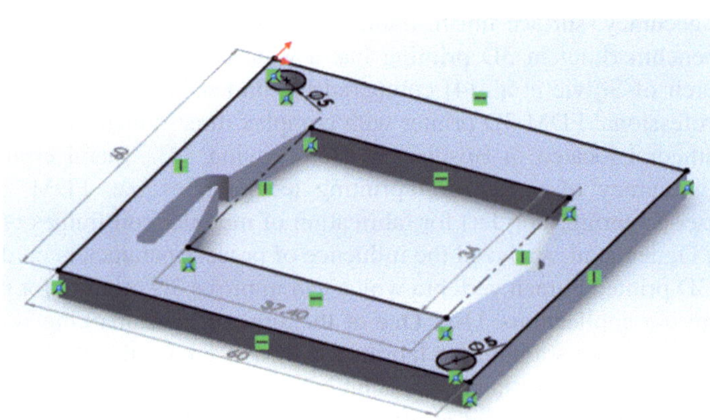

Fig. 4. Digital CAD model of reformer PEM fuel cell middle plate.

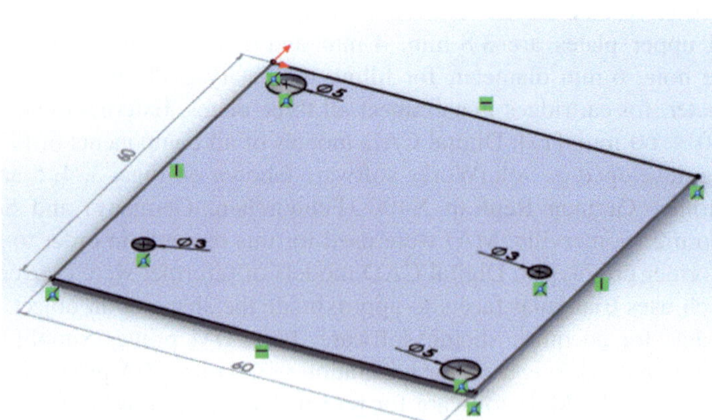

Fig. 5. Digital CAD model of reformer PEM fuel cell upper plate.

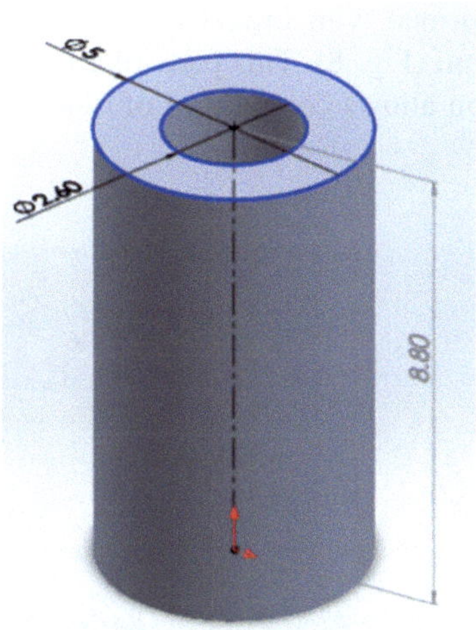

Fig. 6. Digital CAD model of reformer PEM fuel cell cartridge.

Three prototypes of PEM fuel cell reformer were made with a different height resolution of print: 0.3 mm, 0.2 mm and 0.1 mm. 0.3 mm and 0.1 mm height resolutions represent lowest and highest resolution for selected FDM 3D printer. Other printing parameters are the same for all three printing processes. FDM printer is equipped with 0.4 mm brass nozzle, nozzle temperature is set at 200 °C. Platform temperature is set at 60 °C, keeping extruded material on the platform. Hexagonal infill pattern with 50% infill is for all three prototypes. Printing speed is 40 mm/s. Print with the height resolution of 0.3 mm is shown on Fig. 7.

Fig. 7. FDM printing of PEM fuel cell reformer.

As a general rule for FDM printing, parts should be printed with their shortest dimensions in the vertical direction (FDM), [10]. Before the printing of models, first few layers are printed as a stand (raft) for better grip of a prototype to the platform.

SLA printing was performed with lowest height resolution possible for selected SLA printer, which is 0.1 mm (Fig. 8). This particular print combined with FDM print of the same height resolution allows comparison of dimensional accuracy of these 3D printing technologies.

Fig. 8. SLA print of PEM fuel cell reformer.

Larger support structure for SLA printing is needed in order to successfully finish print. Otherwise, print might fall off the platform, resulting in failed process. Every model is printed with a stand for better grip to build platform. All printed prototypes for this research are shown in Fig. 9.

Fig. 9. All printed prototypes of PEM fuel cell reformer: (left to right) FDM 0.3 mm, FDM 0.2 mm, FDM 0.1 mm, SLA 0.1 mm.

SLA print has the smoothest surface, with no layer lines noticeable. FDM print with lowest height resolution has the roughest surface of the three printed by FDM printing technology. The smoothest surface is on the prototype with 0.1 mm resolution, but still not comparable to SLA print.

3 Results and Discussion

Dimensions of interest of all FDM and SLA printed PEM fuel cell reformer components are listed in Tables 1, 2, 3 and 4. Relative deviations of printed models from the digital CAD model are calculated by the equation, [3]:

$$\varepsilon = \frac{l - l_0}{l_0} \tag{1}$$

where l is a measured dimension of printed model and l_0 is the appropriate dimension of digital CAD model. Multiplying equation by 100 gives a percentage value (%) of relative deviation. In Tables 1, 2, 3 and 4, relative deviations are listed in brackets.

Measurements are performed with the digital calliper "ORION 31170210". Dimensional range is 0–150 mm with 0.01 mm resolution. Digital calliper was last time calibrated on 22.03.2018 according to DIN 862:2005 standard in accredited metrology laboratory "21. MAJ DOO" (Belgrade, Serbia), under certificate number 789/2018.

Table 1. 3D printed upper plate

Features	Original CAD model	FDM 0.1 mm res.	FDM 0.2 mm res.	FDM 0.3 mm res.	SLA 0.1 mm res.
External dimension	60 × 60 mm	59.89 × 59.95 mm (−0.002 × −0.001)	59.76 × 59.84 mm (−0.004 × −0.003)	59.70 × 59.82 mm (−0.005 × −0.003)	59.97 × 60.32 mm (−0.0005 × 0.0005)
Two channel holes	5 mm	4.93 mm (−0.014)	4.76 mm (−0.048)	4.69 mm (−0.062)	4.98 mm (−0.004)
	5 mm	4.85 mm (−0.03)	4.72 mm (−0.056)	4.65 mm (−0.07)	5 mm (0)
Inlet and outlet holes	3 mm	2.56 mm (−0.14)	2.51 mm (−0.16)	2.43 mm (−0.19)	2.99 mm (−0.003)
	3 mm	2.78 mm (−0.07)	2.55 mm (−0.15)	2.51 mm (−0.16)	3 mm (0)
Thickness	1 mm	1.18–1.28 mm (0.18–0.28)	1.30–1.44 mm (0.3–0.44)	1.23–1.38 mm (0.23–0.38)	0.93–1 mm (−0.07–0)

For FDM prints, negative values of relative deviation in Tables 1, 2 and 3 suggest material shrinking after a finished print process. Plates printed with 0.3 mm height resolution have maximal external dimension relative deviation of −0.005, i.e. 0.5% shrinkage from original CAD model. Change to 0.2 mm height resolution results in slight improvement in dimensional accuracy of external dimension on all three plates, except in one external dimension in the case of the lower plate, where maximal relative deviation is −0.005 (−0.5%). Best results were measured on plates with 0.1 mm height resolution. Maximal relative deviation was measured on the lower plate with the value of −0.003 (−0.3%), and on middle and upper plate smallest relative deviation is just −0.001 (−0.1%).

Table 2. 3D printed middle plate

Features	Original CAD model	FDM 0.1 mm res.	FDM 0.2 mm res.	FDM 0.3 mm res.	SLA 0.1 mm res.
External dimension	60 × 60 mm	59.87 × 59.91 mm (−0.002 × −0.001)	59.79 × 59.87 mm (−0.003 × −0.002)	59.79 × 59.87 mm (−0.003 × −0.002)	60.00 × 60.00 mm (0 × 0)
Two channel holes	5 mm	4.43 mm (−0.114)	4.14 mm (−0.172)	4.03 mm (−0.194)	5 mm (0)
	5 mm	4.31 mm (−0.138)	4.19 mm (−0.162)	4.14 mm (−0.172)	5 mm (0)
Reaction volume	34 × 37.4 mm	33.90 × 37.46 mm (−0.003 × 0.002)	33.85 × 37.52 mm (−0.004 × 0.003)	33.73 × 37.51 mm (−0.007 × 0.003)	34.02 × 37.4 mm (0.0005 × 0)
Thickness	4 mm	4.64–4.72 mm (0.16–0.18)	4.56–4.80 mm (0.14–0.2)	4.71–4.95 mm (0.18–0.24)	3.91–4 mm (0.02–0)

Table 3. 3D printed lower plate

Features	Original CAD model	FDM 0.1 mm res.	FDM 0.2 mm res.	FDM 0.3 mm res.	SLA 0.1 mm res.
External dimension	60 × 60 mm	59.80 × 59.87 mm (−0.003 × −0.002)	59.72 × 59.73 mm (−0.005 × −0.004)	59.73 × 59.73 mm (−0.005 × −0.005)	60.00 × 60.33 mm (0 × 0.005)
Two channel holes	5 mm	4.69 mm (−0.062)	4.45 × mm (−0.11)	4.44 mm (−0.112)	5 mm (0)
	5 mm	4.65 mm (−0.07)	4.53 mm (−0.094)	4.43 mm (−0.114)	5 mm (0)
Catalyst hole	6 mm	5.83 mm (−0.028)	5.75 mm (−0.042)	5.70 mm (−0.05)	6 mm (0)
Thickness	3.8 mm	4.18–4.27 mm (0.1–0.12)	4.15–4.31 mm (0.09–0.13)	4.34–4.54 mm (0.14–0.19)	3.79–3.8 mm (−0.002–0)

Holes have a greater relative deviation from the original CAD model, than the external dimension of plates. Holes with 5 mm diameter have relative deviation in range from −0.03 (−3%), measured on the upper plate with 0.1 mm height resolution, to −0.194 (−19.4%), measured on the middle plate with 0.3 mm height resolution. Inlet and outlet holes on the upper plate, with 3 mm diameter, have relative deviation from −0.07 (−7%) to −0.19 (−19%), but with more measured dimensions closer to the maximal relative deviation. This suggests that model accuracy depends on the size of the printed features. Smaller features tend to have greater relative deviations.

The same goes for the plate thickness. Largest relative deviation was measured on the upper plate, which has the smallest thickness of the three. The maximal relative deviation value is 0.38 (38%). Middle and lower plates have smaller maximal relative deviation of thickness, i.e. 0.24 (24%) and 0.19 (19%), respectively. All mentioned maximal relative deviations in thickness were measured on plates printed with 0.3 mm height resolution. Overall, largest variation in plate thickness is located at plates printed with 0.3 mm layer height. Larger variation in thickness suggests low flatness of particular plate. Values of thickness in Tables 1, 2 and 3 show that height resolution greatly influences on flatness and straightness of a printed model. On the upper plate thickness varies from 0.18 (18%) to 0.28 (28%), i.e. variation of which is 10%, in the case of the plate printed with 0.1 mm height resolution. For 0.2 mm plate that variation is set at 14% and for 0.3 mm plate at 15%. Best improvement of plate flatness and straightness is present at prints with 0.1 mm height resolution. The same follows for the other two plates.

In Table 4 are listed dimensions for two cartridges for each print. Because of the difficulty of printing circular features layer-by-layer, cartridges had to be printed with the longest dimension in a vertical direction.

Table 4. 3D printed cartridge

Features	Original CAD model	FDM 0.1 mm res.	FDM 0.2 mm res.	FDM 0.3 mm res.	SLA 0.1 mm res.
Height	8.8 mm	9.00 mm (0.023)	9.03 mm (0.026)	9.06 mm (0.029)	8.87 mm (0.008)
	8.8 mm	9.02 mm (0.025)	9.06 mm (0.029)	9.07 mm (0.030)	8.76 mm (−0.004)
Ex. diameter	5 mm	4.92 mm (−0.016)	4.87 mm (−0.026)	5.12 mm (0.024)	4.96 mm (−0.008)
	5 mm	4.94 mm (−0.012)	4.83 mm (−0.034)	5.13 mm (0.65)	4.97 mm (−0.006)
In. diameter	2.6 mm	2.40 mm (−0.077)	2.24 mm (−0.138)	1.90 mm (−0.269)	2.6 mm (0)
	2.6 mm	2.39 mm (−0.081)	2.29 mm (−0.119)	2.14 mm (−0.177)	2.6 mm (0)

Tables 1, 2, 3 and 4 show undoubtedly that shrinkage of FDM printed models are present in a horizontal plane, and that in height FDM material expands. SLA print with the layer resolution of 0.1 mm (Tables 1, 2, 3 and 4) has noticeably higher dimensional accuracy than FDM print of the same height resolution. Some dimensions are equal to corresponding dimensions of digital CAD model, e.g. a diameter of holes and the external dimension of plates printed closer to build platform. Deviations from digital CAD model are present in height of cartridges and the external dimension of plates positioned vertically to build platform during the printing process. This suggests that dimensional inaccuracy of SLA prints originates from an accumulation of errors during the print process. Highest deviations from original model are present in last layers of print, i.e. highest accuracy is present in dimensions closer to build platform.

For the purpose of better understanding the behaviour of polymeric materials used in 3D printing technologies and to obtain higher accuracy results, future research will include Digital Image Correlation (DIC) method of receiving relative movement, i.e. deformation, of the observed structure, [11–14, 18–21].

4 Conclusion

3D printing represents an emerging manufacturing technology, with plenty of space for research and development. One of the biggest challenges in 3D printing is to minimize dimensional inaccuracies compared to digital CAD model, caused by 3D printing process and material properties. For research geometrical benchmarking is used to measure and compare the accuracy of different printing technologies and printing parameters within the same 3D printing technology.

FDM and SLA nowadays present two most common 3D printing technologies. In this research, a PEM fuel cell reformer is used for geometric benchmarking to compare dimensional accuracy, surface finish, flatness and straightness of printed models of two different printing technologies. The aim of benchmarking is to compare strengths and weaknesses of printing parameters and different printing technologies in order to make improvements.

Acknowledgement. This research is carried out under the NATO SPS Project EAP.SFPP 984738, and the Ministry of Education and Science of the Republic of Serbia projects TR35040, TR35006 and III 43007.

References

1. ASTM F2792-12a: Standard Terminology for Additive Manufacturing Technologies. doi: https://doi.org/10.1520/f2792-12a
2. Shah, P., Racasan, R., Bills, P.: Comparison of different additive manufacturing methods using computed tomography. Case Stud. Nondestruct. Test. Eval. (2016). https://doi.org/10.1016/j.csndt.2016.05.008
3. Yankov, E., Nikolova, M.P.: Comparison of the accuracy of 3D printed prototypes using the stereolithography (SLA) method with the digital CAD models. In: MATEC Web of Conferences, vol. 137, p. 02014 (2017). https://doi.org/10.1051/matecconf/201713702014
4. Sljivic, M., Pavlovic, A., Ilic, J., Stanojevic, M., Todorovic, S.: Comparing the accuracy of professional and consumer grade 3D printers in complex models production. FME Trans. **45**, 348–353 (2016). https://doi.org/10.5937/fmet1703348S
5. Lotric, A., Sekavcnik, M., Pohar, A., Likozar, B., Hocevar, S.: Conceptual design of an integrated thermally self-sustained methanol steam reformer: high-temperature PEM fuel cell stack manportable power generator. Int. J. Hydrogen Energy **42**, 16700–16713 (2017). https://doi.org/10.1016/j.ijhydene.2017.05.057
6. Fahad, M., Hopkinson, N.: A new Benchmarking part for evaluating the accuracy and repeatability of additive manufacturing (AM) processes. In: Proceedings of the ICMPAE 2012 – 2nd International Conference on Mechanical, Production and Automobile Engineering (2012)
7. Cicek, Y., Altinkaynak, A., Balta, E.: Numerical and experimental analysis of infill rate on the mechanical properties of fused deposition modelling polylactic acid parts. In: Proceedings of the SPE ANTEC® Anaheim (2017)
8. Alafaghani, A., Qattawi, A., Alrawi, B., Guzman, A.: Experimental optimization of fused deposition modelling processing parameters: a design-for-manufacturing approach. In: Proceedings of the NAMRC 45 – 45th SME North American Manufacturing Research Conference (2017)

9. Ibrahim, M., Hafsa, M.N.: Dimensional accuracy of additive manufacturing model with different internal structure for investment casting implementation. In: International Integrated Engineering Summit (IIES 2014), 1–4 December 2014. Universiti Tun Hussein Onn Malaysia, Johor (2014)
10. McCoy, C., Farbman, D.: Materials testing of 3D printed ABS and PLA samples to guide mechanical design. In: Proceedings of the ASME 2016 International Manufacturing Science and Engineering Conference (2016)
11. Milosevic, M., Miletic, V., Mitrovic, N., Manojlovic, D., Savic-Stankovic, T., Maneski, T.: Measurement of local deformation fields in dental composites using 3D optical system. Chemicke Listy **105**, 751–753 (2011). ISSN: 0009-2770
12. Miletic, V., Manojlovic, D., Milosevic, M., Mitrovic, N., Savic, Stankovic T., Maneski, T.: Analysis of local shrinkage patterns of self-adhering and flowable composites using 3D digital image correlation. Quintessence Int. **42**(9), 797–804 (2011). ISSN (print) 0033-6572; ISSN (online) 1936-7163 (2011)
13. Manojlovic, D., Dramicanin, M.D., Milosevic, M., Zekovic, I., Cvijovic-Alagic, I., Mitrovic, N., Miletic, V.: Effects of a low-shrinkage methacrylate monomer and monoacylphosphine oxide photoinitiator on curing efficiency and mechanical properties of experimental resin-based composites. Mater. Sci. Eng. C **58**, 487–494 (2016)
14. Lezaja, M., Veljovic, Đ., Manojlovic, D., Milosevic, M., Mitrovic, N., Janackovic, Đ., Miletic, V.: Bond strength of restorative materials to hydroxyapatite inserts and dimensional changes of insert-containing restoration during polymerization. Dental Mater. **31**(2), 171–181 (2015). https://doi.org/10.1016/j.dental.2014.11.017
15. Queral, V., Rincon, E., Mirones, V., Rios, L., Cabrera, S.: Dimensional accuracy of additively manufactured structures for modular coil windings of stellarators. Fus. Eng. Des. **124**, 173–178 (2017). https://doi.org/10.1016/j.fusengdes.2016.12.014
16. Ogden, K.M., Aslan, C., Ordway, N., Diallo, D.: Factors affecting dimensional accuracy of 3-D printed anatomical structures derived from CT data. J. Digit. Imaging **28**(6), 654–663 (2015). https://doi.org/10.1007/s10278-015-9803-7
17. Milosevic, M., Mladenovic, G., Sedmak, A., Plohar, A., Likozar, B.: Design and manufacture of reformer in polymer electrolyte membrane fuel cell. Struct. Integr. Life **17** (1), 21–24 (2017). ISSN: 1451-3749
18. Sedmak, A., Milosevic, M., Mitrovic, N., Petrovic, A., Maneski, T.: Digital image correlation in experimental mechanical analysis. Struct. Integr. Life **12**(1), 39–42 (2012). ISSN: 1451-3749
19. Miletic, V., Peric, D., Milosevic, M., Manojlovic, D., Mitrovic, N.: Local deformation fields and marginal integrity of sculptable bulk-fill, low-shrinkage and conventional composites. Dental Mater. **32**(11), 1441–1451 (2016). https://doi.org/10.1016/j.dental.2016.09.011
20. Colic, K., Sedmak, A., Legweel, K., Milosevic, M., Mitrovic, N., Miskovic, Z., Hloch, S.: Experimental and numerical research of mechanical behaviour of titanium alloy hip implant. Tech. Gaz. **24**(3), 709–713 (2017). https://doi.org/10.17559/tv-20160219132016
21. Tatic, U., Colic, K., Sedmak, A., Miskovic, Z., Petrovic, A.: Evaluation of the locking compression plates stress-strain fields. Tech. Gaz. **25**(1), 112–117 (2018). https://doi.org/10.17559/TV-20170420121538

Application of Numerical Methods in Design and Analysis of Orthopedic Implant Integrity

Katarina Čolić[1(✉)], Aleksandar Grbović[2], Aleksandar Sedmak[2], and Kaled Legweel[3]

[1] Innovation Center of Faculty of Mechanical Engineering, University of Belgrade, Belgrade, Serbia
kbojic@mas.bg.ac.rs
[2] Faculty of Mechanical Engineering, University of Belgrade, Belgrade, Serbia
[3] Sebha High Institute, Sebha University, Sabha, Libya

Abstract. In this paper a numerical analysis of hip implant model and hip implant model with a crack in a biomaterial is presented. Hip implants still exhibit problem of premature failure, promoting their integrity and life at the top of the list of problems to be solved in near future. Any damage due to wear or corrosion is ideal location for crack initiation and further fatigue growth. Therefore, this paper is focused on integrity of hip implants with an aim to improve their performance and reliability. Numerical models are based on the finite element method (FEM), including the extended FEM (X-FEM). FEM became a powerful and reliable numerical tool for analysis of structures subjected to different types of load in cases where solving of these problems was too complex for exclusively analytical methods. FEM is a method based on discretization of complex geometrical domains into much smaller and simpler ones, wherein field variables can be interpolated using shape functions. Numerical analysis was performed on three- dimensional models, to investigate mechanical behaviour of a hip implant at acting forces from 3.5 to 6.0 kN. Short theoretical background on the stress intensity factors computation is presented. Results presented in this paper indicate that acting forces can lead to implant failure due to stress field changes. For the simulation of crack propagation extended finite element method (XFEM) was used as one of the most advanced modelling techniques for this type of problem.

Keywords: Hip replacement implant · Stress intensity factor · Crack growth Numerical simulations · Extended finite element method

1 Introduction

In the field of orthopaedic surgery, it is very important to determine the stress-strain condition of the implants and bones. Mechanical loads of organs and tissues within a human body cause stresses and strains. Strain caused by mechanical load produce internal mechanical forces within the body. For simpler geometric shapes, analysis of mechanical behaviour can be determined analytically, wherein for complex cases, it is necessary to apply a numerical method, such as finite element method (FEM). Finite element analysis, represents a numerical analysis used for solving of complex geometry

N. Mitrovic et al. (Eds.): CNNTech 2018, LNNS 54, pp. 96–111, 2019.
https://doi.org/10.1007/978-3-319-99620-2_8

problems, for which obtaining of an analytical solution is extremely difficult. FEM is a method based on discretization of complex geometrical domains into much smaller and simpler ones, wherein field variables can be interpolated using shape functions.

Application of FEM in biomedical applications is becoming more and more frequent, causing complications in terms of complicated geometry, when it is almost impossible to reach an analytical solution. During previous studies, this method was used to determine the stress state of the hip implants, as well as on orthopedic plates of different geometries [1–3].

In FEM, a complex region which defines the continuum is discretized into simple geometric shapes - elements. It is assumed that these elements have properties and relations which can be mathematically expressed as unknown quantities in certain points of elements - nodes. A process of connecting and combining of individual elements in a given system is applied. After taking into account the influence of load and boundary conditions, a series of linear or non-linear equations is typically obtained. Solving of these equations provides an approximate behaviour of a continuum or a system. The algorithm for this method consists of the following steps: continuum discretization, selection of interpolation functions, calculating of system properties, forming of algebraic equations, solving of algebraic equation systems, and calculation of necessary influences in nodes for individual finite elements.

The advantages of applying FEM include: it is applicable to complex geometries, complex types of analysis, complex loads, and models made of non-homogeneous materials, etc. Types of errors that can occur include discretization errors, as well as formulation and numerical errors. The basic equation of FEM for static load conditions is $\{F\} = [K] \times \{u\}$, where $[K]$ is the general or global stiffness matrix, $\{u\}$ is the global displacement vector [4, 5].

FEM saw its first use in orthopaedic biomechanics in 1972, for the purpose of assessing of stresses in human bones. Since then, this method has been applied with increasing frequency in stress state analysis of bones and prosthetics, as well as fracture fixation, [1–3, 6–15], an example model is presented in Fig. 1 [1].

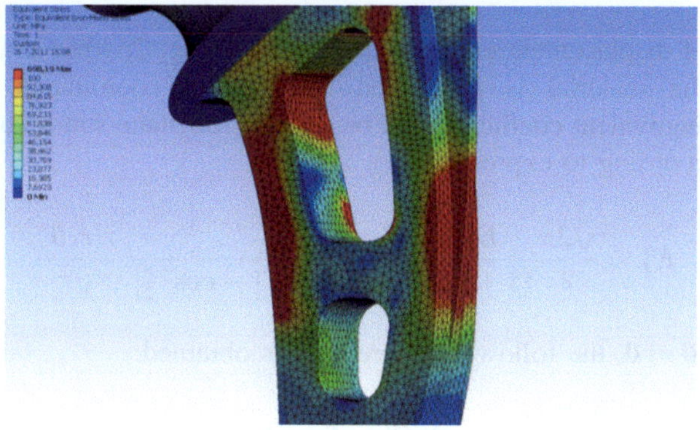

Fig. 1. Stress distribution on a hip implant obtained by the FEM, [1]

In addition to bones, this method can be used in analysis of numerous other tissues and organs. In case of orthopaedics, there has always been a significant interest in stress and load. However, mathematical tools available for stress analysis in classic mechanics were not suitable for calculations of extremely irregular structural properties of bone. Hence, the use of FEM represented a logical step due to its unique ability to determine stress state in structures with complex shape, load and material behaviour, [14–19]. In this paper, the basics of finite element method and its application to prosthetics will be presented, with particular attention to modelling of a hip replacement prosthesis.

2 Application of FEM in Analysis of Behavior of Biomaterials with Cracks

Modelling of fracture mechanics problems requires an adequate treatment of displacement and stress field singularity around the crack tip, where the biggest problem is reflected in drastic increase in discretization error, which occurs when using classic FE, such as the eight-node element. The most efficient solution is obtained by using the reduction technique (reducing the error to only 1%) or by applying special FE around the crack tip, which contains the strain field singularity [4, 5, 20–23].

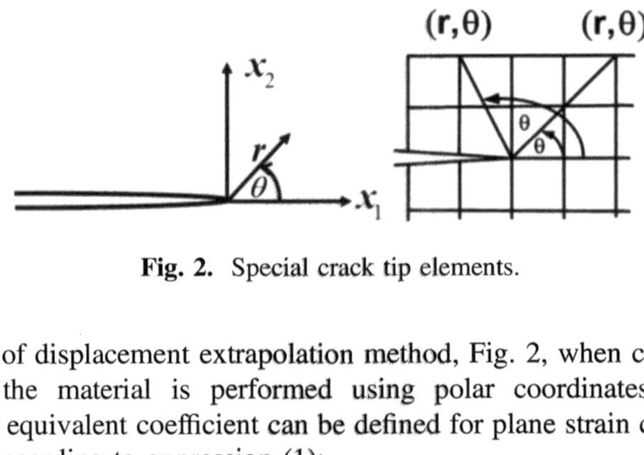

Fig. 2. Special crack tip elements.

In the case of displacement extrapolation method, Fig. 2, when crack propagation simulation in the material is performed using polar coordinates, the following expressions for equivalent coefficient can be defined for plane strain conditions, and it is determined according to expression (1):

$$K_I^* = \frac{\sqrt{2\pi}}{2} \frac{E}{(1+\nu)} \frac{1}{\sin\frac{\theta}{2}\left[2(1-\nu) - \cos^2\frac{\theta}{2}\right]} \frac{v(r,\theta)}{\sqrt{r}} \tag{1}$$

If the angle $\theta = 0$, the following expression is obtained:

$$K_I^* = \frac{\sqrt{2\pi}}{4} \frac{E}{1-\nu^2} \frac{v(r,\pi)}{\sqrt{r}} \tag{2}$$

$$K_I = \lim_{r \to \infty} K_I^* \tag{3}$$

Fracture mechanics parameters can be determined in a number of different methods, such us displacement extrapolation, J-Integral, stiffness derivative method, etc.

3 Application of X-FEM Method to Crack Growth Simulation

In order to evaluate the influence of initial defects in material on strength and life of structures, finite element analysis is applied to cracks of various shapes, sizes and locations. In these analyses, FEM is limited, since changes in crack topology require additional generating of mesh domain. This represents a significant constraint and complicates crack growth simulation on complex geometries. Extended finite element method (X-FEM) was developed in order to make calculations easier, which was required during positioning of arbitrary cracks within a finite element model [20, 24]. It is a very wide field of application of this method, first of all in determining the behavior of material with crack in the engineering, as well as in biomedical applications [25–27].

XFEM uses enhancement functions as a means of displaying all forms of discontinuous behaviour, such as crack displacement. Enhancement functions are introduced into the displacement approximation for only a small number of finite elements, relative to the size of the whole domain. Additional degrees of freedom are introduced for all elements where the discontinuity is present, and in some cases - depending on the type of the selected function - into adjacent elements, which are then referred to as mixed elements [28–31].

Displacement approximation can be expressed in the following way, by applying enhancement functions:

$$\boldsymbol{u}^h(\boldsymbol{x}) = \sum_i N_i(\boldsymbol{x}) \left[\boldsymbol{u}_i + \sum_j v^j(\boldsymbol{x}) \boldsymbol{a}_i^j \right] \tag{4}$$

The unity property is based on the fact that the sum of interpolation functions of finite elements equals one. Assuming that the unity property is fulfilled, additional enriching functions, i.e. improvement functions, can be given in displacement approximation. In this case, application of standard X-FEM displacement formulation approximates displacements as:

$$\begin{aligned} u(\xi,\eta,\zeta) = \sum\nolimits_i N_i(\xi,\eta,\zeta) U_i + \sum\nolimits_i N_i(\xi,\eta,\zeta) H(\xi,\eta,\zeta) b_i \\ + \sum\nolimits_i N_i(\xi,\eta,\zeta) \left(\sum\nolimits_j \Psi_j(r,\theta) c_{ji} \right) \end{aligned} \tag{5}$$

where $N_i(\xi,\eta,\zeta)$ are shape functions, $U_i \in \mathbf{R}^3$ are node displacement parameters for all nodes of a hexahedron element: $1 \sim 8$, $b_i \in \mathbf{R}^3$ are parameters of jump function on jump nodes, and $c_{ji} \in \mathbf{R}^3 \times \mathbf{R}^4$ are parameters of the branching function for nodes at the crack tip [28–31].

It is necessary during calculation to determine which mesh elements were divided by the crack and in which element the crack tip is located, taking into account that

X-FEM does not approximate the entire domain. In this sense, an unequivocal identification of elements uses two functions on the level of sets (LS functions), which are based on level set (LS) method.

Jump function H is defined as the sign of the level set φ:

$$H(\xi,\eta,\zeta) = \begin{cases} +1 : \varphi(\xi,\eta,\zeta) > 0 \\ -1 : \varphi(\xi,\eta,\zeta) < 0 \\ \pm 1 : \varphi(\xi,\eta,\zeta) = 0 \end{cases} \tag{6}$$

It should be noticed that function $H(\xi,\eta,\zeta)$ is not well defined when $\varphi(\xi,\eta,\zeta) = 0$, $H(\xi,\eta,\zeta) = \pm 1$ and $[\![H(\xi,\eta,\zeta)]\!] = 2$ merely represents a suitable way of calculating of the jump function in points which are located at the crack surface [28–31].

4 Numerical Calculation of Selected Hip Implant Models

Numerical models were made of a hip prosthesis in order to analyse material behaviour of an implant during load in an ideal case, and also when there is a crack in the material. In this sense, simplifications of problems related to implanting of the prosthetic were performed in order to fulfil the requirements in terms of size. In a realistic case, there are many factors that influence the integrity of the prosthetic, such as state of bones, effects of corrosion and biocompatibility of selected metallic biomaterials [32–35], as well as bone cement properties [36, 37], but it is not possible to simulate all of these effects, since they depend on individual cases as well [38].

For FEM analyses, performed in commercial software ABAQUS, three-dimensional models of the prosthetic and stem were made based on real prosthetic components, one example presented in a Fig. 3.

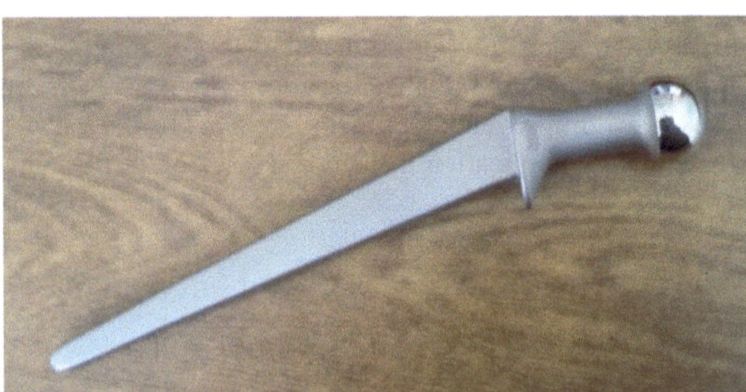

Fig. 3. Hip implant model made of CoCrMo alloy.

Selected model is made of CoCrMo alloy [33, 39], whose properties are given in Table 1.

Table 1. Mechanical properties

Properties	CoCrMo
Tensile strength	655
Yield point (0.2%), MPa	450
Elongation, %	8
Reduction area, %	8
Fracture point, $MPa\sqrt{m}$	310

4.1 Development of Numerical Models

The geometry of the stem has a significant effect on prosthesis performance. Stem with a smooth surface, generally speaking, reduces the stress concentration and enable significant fatigue resistance. Stem with a sharp or rugged surface enables good connection and prevents potential sliding of the joint. Level of stress concentration and tendency to fatigue fracture depend on the roughness of the stem surface.

A dimension of adopted geometrical model is given in the Fig. 4.

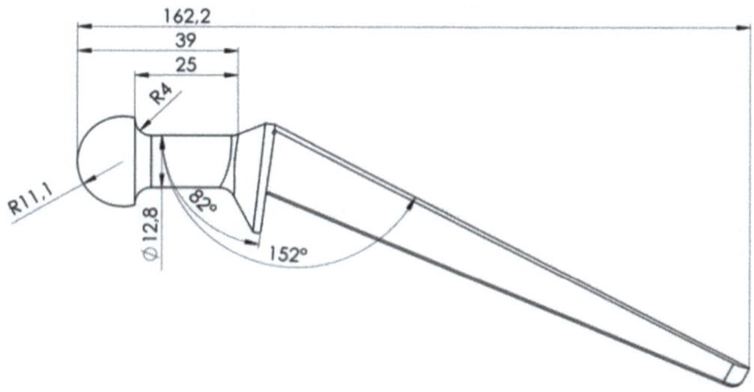

Fig. 4. Hip implant model – dimensions

The constitutive relation which is chosen for the given problem requires the specification of two constants, Young's elasticity modulus and Poisson's ratio.

Values of coefficients used for this problem in an example presented here, are given in Table 2.

Table 2. Values of coefficients used

Material	Elasticity modulus (GPa)	Poisson's ratio
CoCrMo	234	0.3

FEM analysis requires a numerical description of all external loads affecting the structure (points in which they act, magnitude, direction). These loads are usually variable and are not always precisely defined, thus when using FEM analysis, a frequent question is which approach to use in order to obtain useful information.

Certain approximations have been introduced regarding boundary conditions and loads of the model. Taking into account that only the behaviour of metal structures is analysed in terms of crack presence in the biomaterial, and based on the review of relevant literature in terms of the expected location of crack initiation, it was possible to introduce a fixed support approximation for the load bearing structure of the implant in the lower part which is in contact with the bone.

Real load which acts on the implant was introduced in two ways, as compressive load on prosthetic cup, and as a force which acts in a single point on the acetabular part of the prosthetic. Loads defined in this way actually represent an approximation of real hip implant behaviour. Since real load acting on the hip is highest in the simulated direction, and in the first approximation of the bone-implant connection, a completely rigid connection can be assumed.

Two characteristic load types during the walking stage were selected, shown in Table 3, for normal and fast walking. In that sense, the values of the load range from minimum 4.9 BW to maximum 7.6 BW loads. For the purpose of numerical calculation, a person with a weight of 80 kg was selected, thus the expected loads on the implant during normal walking were 3845.5 N and for fast walking 5964.5 N respectively.

Table 3. The maximum force on the hip joint

Activity	The maximum force on the hip joint (a multiple of body weight)
Walking on flat surface	
Slow	4.9
Normal	4.9
Fast	7.6
Climbing the stairs	7.2
Coming down the stairs	7.1
Climbing the steep surface	5.9
Coming down the steep surface	5.1

4.2 Discretization of Structure in Numerical Models

For the purpose of FEM analysis, two types of finite elements were selected, which adequately describe the behaviour of complex three-dimensional structures, such as hip implants. Within the calculation, standard 3D stress elements type libraries were used, which are a part of the applied software package. Applied to all models were two types of 3D elements, an 8-node linear hex type element, with reduced integration and a 10-node tetrahedral type of quadratic element.

Discretization parameters for model calculation are shown in Table 4, and include the selected shape and number of elements and nodes.

Table 4. Discretization parameters of a numerical model

Model
Implant base:
Hexahedral type:(C3D8R) 8-node linear Reduced integration
Implant cup:
Hexahedral type:(C3D8R) 8-node linear Reduced integration *Tetrahedral type*: (C3D10) 10-node quadratic
Total number of elements: 7 253
Number of nodes: 9 708

The finite elements mesh generated on the calculation modelis shown in Fig. 5.

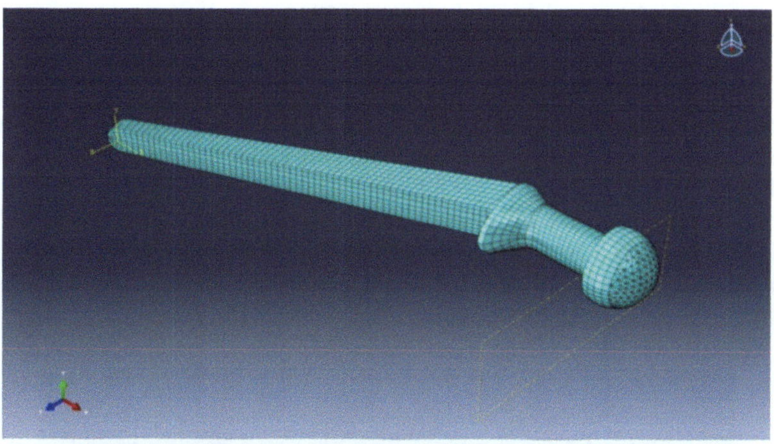

Fig. 5. Numerical model - FE mesh

5 Development of a Model with a Crack

The most recent method which is applied to modelling of cracked material behaviour is based on extended finite element method, X-FEM. [24–27].

Based on literature analysis, it was assumed that in places where stem was connected to the acetabular part, wear or corrosion can occur in material, and as a result of that, material damage may occur, i.e. the appearance of cracks [33, 40–42].

During further model development and preparation for FEM analysis, generating of the mesh on the implant model was performed, with an initial crack and a tendency to obtain a mesh as fine as possible.

Shown in Fig. 6 is the implant model with a generated finite elements mesh and an initial crack positioned in the critical area.

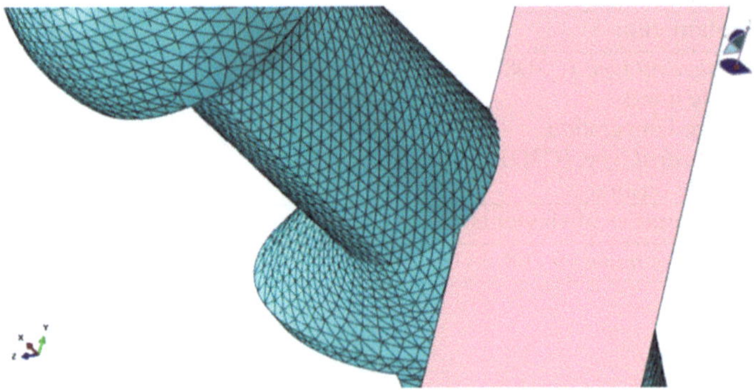

Fig. 6. Display of a discretized implant model with a crack in the critical area

6 Results of Numerical Analysis

To investigate the difference between results for loads defined by standard defined and real loads on implants that can appear in practice, it is necessary to analyze the prosthesis under a body weight static loading, and under the maximum load that can occur during the walking cycle.

For numerical model the distribution of von Mises stressis shown in Fig. 7, while applied loads is 3845.5 N on the selected implant surface that is 422 mm^2. For this model a standard element library is used, and the model is made using the C3D8R and

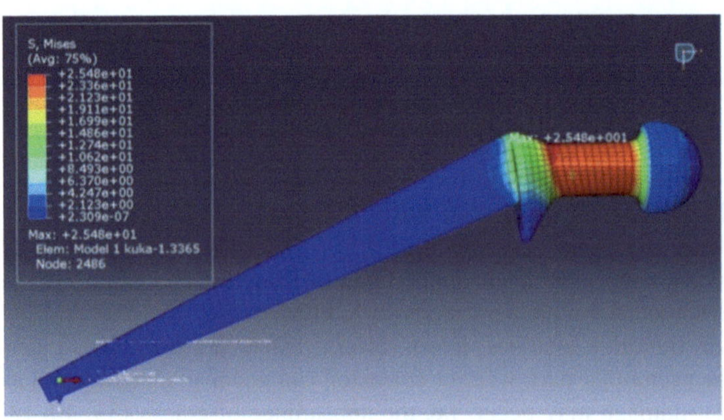

Fig. 7. Von Mises stress distributionon a numerical model.

C3D10 element types. Total number of elements is 7.253 and total number of nodes is 9.708. For this analysis total CPU time is 7.300 and deformation scale factor for graphical representation is 500.

Based on the stress field analysis for the characteristic loads, maximum stress values on an implant model are obtained, with two different types of load structure presented in this paper. The results are shown in Table 5.

Table 5. Results of numerical analysis

Activity	Load on implant (N)	Von Mises stress value (MPa)
Normal walking	3845.5	254
Fast walking	5964.5	387

Referring to the analysis of stress and strain state on the numerical model, it can be concluded that the area of maximum stress and strain values is right at the expected places on implant geometry. Numerical simulations on model show that stress concentration occurs in precisely those areas where in practice fatigue fracture occurred in implants of similar or same geometry.

With regard to these observations, further numerical analysis included numerical models of an implant with fatigue crack in the biomaterial, set just in spots of the highest stress concentration.

Crack propagation rate da/dN was obtained using Paris crack growth low, Paris exponent n is 2.28, and Paris coefficient is 2.05 10^{-11}.

Applied parameters for the numerical calculation of the fatigue crack propagation in biomaterial of the implant are given in Table 6.

Table 6. Numerical parameters for crack propagation behavior

Implant biomaterial
CoCr superalloy
Crack growth low:
$\frac{da}{dN} = C(\Delta K)^n$
Numerical calculation parameters:
$n = 2.28$
$C = 2.056\ 10^{-11}$
$Y_{ie.} = 1751$
$UTS = 1889$
$K_{Ie} = 3649$
$K_{IC} = 2780$

The numerical analysis were done in numerical package Morpheo, which is based on application of extended finite element method, and is supported by numerical software for simulation and finite element analysis Abaqus.

The initial crack was set on a spot where occurrence of fatigue and the micro pitting it is expected.

Figure 8 shows the critical area on mono-block hip implant models in terms of fatigue crack appearance, i.e. the numerical model with crack inserted and finite element mesh generated, and as well the crack propagation in the material.

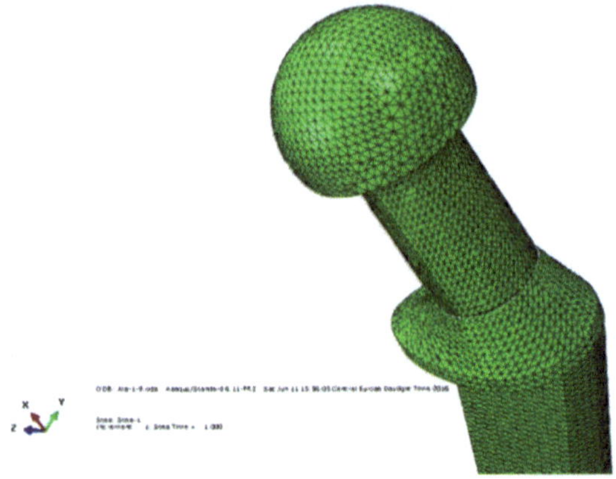

Fig. 8. Crack propagation on a numerical model.

Figure 9 shows the expansion of cracks in the material and stress distribution at the critical crack length for selected implant model.

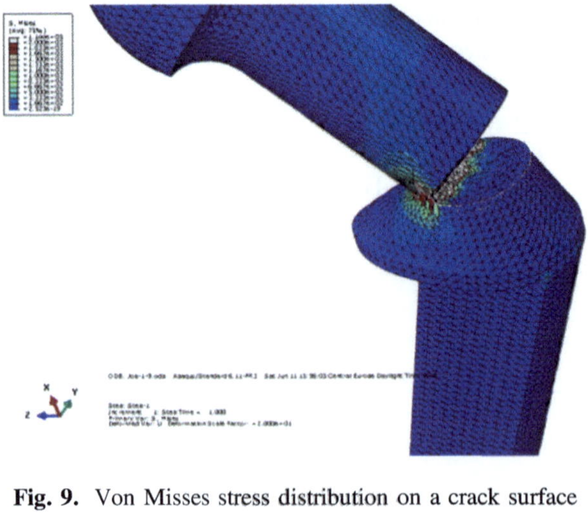

Fig. 9. Von Misses stress distribution on a crack surface

By applying numerical analysis on the model of an implant with crack fracture mechanics parameters were determined, i.e. value of the stress intensity factor K_I, K_{II} K_{III} i K_{ef}. It should be noted that all these values are determined for each calculation step. Based on theoretical considerations of the crack opening modes, it is clear that in this case of loading conditions on an implant, stress intensity factor values are much higher for the mode I than for modes II and III.

For each step of numerical calculation obtained stress intensity factors values K_I K_{II} and K_{III} are presented in Table 7.

Table 7. K_I values for numerical model

Calculation step	Crack length	K $MPa\sqrt{m}$		
		K_I	K_{II}	K_{III}
1	0.5	747.43	38.29	16.85
2	1	887.10	16.17	15.80
3	1.5	1042.62	21.98	7.93
4	2	1288.52	17.01	0.83
5	2.5	1556.34	16.65	6.69
6	3	1983.27	33.55	14.19
7	3.5	2716.15	37.63	22.28
8	4	3698.11	51.62	6.78
9	4.5	5806.34	109.64	33.78
10	5	10,447.68	93.51	1.16
11	5.5	27,257.73	989.75	193.49
12	6	114,211.8	191.97	1148.49

It is recommended to assume an initial crack in the material in accordance with the fracture criteria K_{IC}, when crack growth occurs in case of $K_I \geq K_{IC}$, i.e. the linear elastic analysis fracture criteria.

Obtained values suggest that the hip prosthetic with an initial crack, subjected to a normal walking, i.e. which works under walking cycle normal conditions, will have a number of cycles equal to 29,493 for numerical model before final fracture and prosthesis failure. Results were numerically obtained and show the number of cycles after the crack was initiated, and with the application of numerical simulation, the number of cycles for which the prosthetic functions under normal conditions was determined, assuming there are flaws in the material.

X-FEM had shown that it is an extremely efficient tool for numerical modeling of cracks in LEFM. Compared to standard FEM, X-FEM introduces significant improvements into numerical modeling of crack growth. Main advantages are that the finite element mesh does not need to be adjusted to crack boundaries (crack surface) in order to include geometric discontinuity, and that there is no need to regenerate the mesh in crack growth simulations.

It was shown that by applying modern numerical methods of biomaterial behaviour analysis it is possible to monitor three-dimensional crack behaviour in the material, as well as determining of characteristic fracture mechanics parameters.

7 Conclusions

Generally speaking, as shown and discussed in this paper, there are many influencing factors regarding integrity and life of hip implants. It is necessary to know precisely the loads, including variable stresses due to dynamic loading, biomaterial properties and corrosion behaviour, including fracture mechanics properties, and stress distribution focused on concentration areas. Therefore, new, modern methods are needed for thorough analysis of this significant problem, starting from advanced experimental methods, using fracture mechanics approach to estimate structural integrity, and advance FEM numerical simulations, including fatigue crack growth, both in experimental and numerical research.

(1) FEM is reliable and powerful tool for stress-strain analysis of complex shaped implants, such as the artificial hip. It has been shown that by applying modern numerical methods to biomaterial behaviour analysis, it is possible to monitor three-dimensional crack behaviour in a material, as well as to determine fracture mechanics parameters.
(2) Numerical simulation on developed models show that stress concentration occurs precisely in those areas, where in the examples from the practice there wasa fatigue fracture on the hip implant of very similar geometry. The areas on numerical models with maximum stress concentration coincide with the most common locations for the formation of cracks in biomaterials, which eventually lead to the weakening of the integrity of the prosthesis, or to failure.
(3) A finite element analysis was performed using three-dimensional models to examine the mechanical behaviour of hip prostheses at forces ranging from 3.5 to 6.0 kN. Results show that the force magnitudes acting on the implant are of interest, that according to implant biomaterial and design they can cause implant stress field changes, which can lead to structure integrity problems and implant failure.
(4) Extended finite element method (X-FEM) provided good agreement of experimental and numerical results for the fatigue crack growth.

Orthopaedic biomechanics problems are not examined easily in vivo due to inaccessible locations and invasive methods. In order to obtain clinically useful models, the next step should involve the development of numerical models of the human bones in which the implant is located. It should be noted that the construction of these models is not an easy task, i.e. to get the most accurate results the first step would be scanning the human hip in various patients to form a basis from which the numerical models would be formed. Future work might also include design of numerical models that will take into account the impact of compound between human bones and hip implants, as well as simulation of crack propagation in the cement biomaterial.

References

1. Colic, K., Sedmak, A., Legweel, K., Miloševic, M., Mitrovic, N., Miškovic, Ž., Hloch, S.: Experimental and numerical research of mechanical behaviour of titanium alloy hip implant. Tehnicki vjesnik – Tech. Gaz. **24**(3), 709–713 (2017)
2. Tatić, U., Čolić, K., Sedmak, A., Mišković, Ž., Petrović, A.: Evaluation of the locking compression plates stress-strain fields. Tehničkivjesnik – Technical Gazette **25**(1), 112–117 (2018)
3. Milovanović, A., Sedmak, A., Čolić, K., Tatić, U., Đorđević, B.: Numerical analysis of stress distribution in total hip replacement implant. Struct. Integr. Life **17**(2), 139–144 (2017)
4. Reddy, J.N.: An Introduction to the Finite Element Method. McGraw-Hill, New York (2005)
5. Hutton, D.V.: Fundamentals of Finite Element Analysis. Mc Graw Hill, New York (2004)
6. Senalp, A.Z., Kayabasi, O., Kurtaran, H.: Static, dynamic and fatigue behavior of newly designed stem shapes for hip prosthesis using finite element analysis. Mater. Des. **28**, 1577–1583 (2007)
7. Colic, K., Sedmak, A., Grbovic, A., Tatic, U., Sedmak, S., Djordjevic, B.: Finite element modeling of hip implant static loading. Proc. Eng. **149**, 257–262 (2016)
8. Bennett, D., Goswami, T.: Finite element analysis of hip stem designs. Mater. Des. **29**(1), 45–60 (2008)
9. Pyburn, E., Goswami, T.: Finite element analysis of femoral components paper III-hip joints. Mater. Des. **25**, 705–713 (2004)
10. Paliwal, M., Gordon Allan, D., Filip, P.: Failure analysis of three uncemented titanium-alloy modular total hip stems. Eng. Fail. Anal. **17**(5), 1230–1238 (2010)
11. Gross, S., Abel, E.W.: A finite element analysis of hollow stemmed hip prostheses as a means of reducing stress shielding of the femur. J. Biomech. **34**(8), 995–1003 (2001)
12. Hrubina, M., Horák, Z., Bartoška, R., Navrátil, L., Rosina, J.: Computational modeling in the prediction of Dynamic Hip Screw failure in proximal femoral fractures. J. Appl. Biomed. **11**(3), 143–151 (2013)
13. Balac, I., Colic, K., Milovancevic, M., Uskokovic, P., Zrilic, M.: Modeling of the matrix porosity influence on the elastic properties of particulate biocomposites. FME Trans. **40**(2), 81–86 (2012)
14. Geringer, J., Imbert, L., Kim, K.: Computational Modelling of Biomechanics and Biotribology in the Musculoskeletal System, Computational Modeling of Hip Implants, Biomaterials and Tissues, pp. 389–416. Woodhead Publishing Limited, Sawston (2014)
15. Huiskes, R., Chao, E.Y.S.: A survey of finite element analysis in orthopedic biomechanics. First Decade J. Biomech. **16**(6), 385–409 (1983)
16. Prendergast, P.J.: Clinical finite element models in tissue mechanics and orthopedic implant design. Biomechanics **12**(6), 343–366 (1997)
17. Huiskes, R., Hollister, S.J.: From structure to process, from organ to cell: recent developments of FE-analysis in orthopaedic biomechanics. J. Biomech. Eng. **115**, 520–527 (1993)
18. Yosibasha, Z., Katza, A., Milgromb, C.: Toward verified and validated FE simulations of a femur with a cemented hip prosthesis. Med. Eng. Phys. **35**(7), 978–987 (2013)
19. Marangalou, J.H., Ito, K., van Rietbergen, B.: A new approach to determine the accuracy of morphology–elasticity relationships in continuum FE analyses of human proximal femur. J. Biomech. **45**(16), 2884–2892 (2012)
20. Belytschko, T., Black, T.: Elastic crack growth in finite elements with minimal remeshing. Int. J. Numer. Methods Eng. **45**(5), 601–620 (1998)

21. Moës, N., Dolbow, J., Belytschko, T.: A finite element method for crack growth without remeshing. Int. J. Numer. Methods Eng. **46**(1), 131–150 (1999)
22. Karihaloo, B.L., Xiao, Q.Z.: Modelling of stationary and growing cracks in FE framework without remeshing: a state-of-the-art review. Comput. Struct. **81**(3), 119–129 (2003)
23. Maligno, A.R., Rajaratnam, S., Leen, S.B., Williams, E.J.: A three-dimensional (3D) numerical study of fatigue crack growth using remeshing techniques. Eng. Fract. Mech. **77** (1), 94–111 (2010)
24. Kastratovic, G., Grbovic, A., Vidanovic, N.: Approximate method for stress intensity factors determination in case of multiple site damage. Appl. Math. Model. **39**(19), 6050–6059 (2015)
25. Kredegh, A., Sedmak, A., Grbovic, A., Sedmak, S.: Stringer effect on fatigue crack propagation in A2024-T351 aluminum alloy welded joint. Int. J. Fatigue **105**, 276–282 (2017)
26. Djurdjevic, A., Zivojinovic, D., Grbovic, A., Sedmak, A., Rakin, M., Dascau, H., Kirin, S.: Numerical simulation of fatigue crack propagation in friction stir welded joint made of Al 2024-T351 alloy. Eng. Fail. Anal. **58**, 477–484 (2015)
27. Colic, K., Sedmak, A., Grbovic, A., Burzić, M., Hloch, S., Sedmak, S.: Numerical simulation of fatigue crack growth in hip implants. Proc. Eng. **149**, 229–235 (2016)
28. Belytschko, T., Black, T.: Elastic crack growth in finite elements with minimal remeshing. Int. J. Numer. Methods Eng. **45**(5), 601–620 (1999)
29. Moës, N., Dolbow, J., Belytschko, T.: A finite element method for crack growth without remeshing. Int. J. Numer. Methods Eng. **46**(1), 131–150 (1999)
30. Mohammadi, S.: Extended Finite Element Method for Fracture Analysis of Structure. Blackwell Publishing Ltd., Oxford (2008)
31. Abdelaziz, Y., Hamouine, A.: A survey of the extended finite element. Comput. Struct. **86** (11–12), 1141–1151 (2008)
32. Affatato, S., Colic, K., Hut, I., Mirjanić, D., Pelemis, S., Mitrovic, A.: Short History of Biomaterials Used in Hip Arthroplasty and Their Modern Evolution. Biomaterials in Clinical Practice. Springer, Cham (2018)
33. Sedmak, A., Čolić, K., Burzić, Z., Tadić, S.: Structural integrity assessment of hip implant made of cobalt-chromium multiphase alloy. Struct. Integr. Life **10**(2), 161–164 (2010)
34. Colic, K., Sedmak, A., Gubeljak, N., Burzic, M., Petronic, S.: Experimental analysis of fracture behavior of stainless steel used for biomedical applications. Struct. Integr. Life **12** (1), 59–63 (2012)
35. Cvijović-Alagić, I., Cvijović, Z., Mitrović, S., Panić, V., Rakin, M.: Wear and corrosion behaviour of Ti–13Nb–13Zr and Ti–6Al–4V alloys in simulated physiological solution. Corros. Sci. **53**(2), 796–808 (2011)
36. Hloch, S., Monka, P., Hvizdos, P., Jakubeczyova, D., Kozak, D.V., Colic, K.G., Kl'oc, J., Magurova, D.: Thermal manifestations and nanoindentation of bone cements for orthopaedic surgery. Thermal Sci. **18**(1), S251–S258 (2014)
37. Kühn, K.D.: Bone Cements: Up-To-Date Comparison of Physical and Mechanical Properties of Commercial Materials. Springer, New York (2000)
38. Hernandez-Rodriguez, M.A.L., Ortega-Saenz, J.A., Contreras-Hernandez, G.R.: Failure analysis of a total hip prosthesis implanted in active patient. J. Mech. Behav. Biomed. Mater. **3**, 619–622 (2010)
39. Legweel, K., Sedmak, A., Čolić, K., Burzić, Z., Gubeljak, L.: Elastic-plastic fracture behaviour of multiphase alloy MP35N. Struct. Integr. Life **15**(3), 163–166 (2015)

40. Lee, E.W., Kim, H.T.: Early fatigue failures of cemented, forged, cobalt-chromium femoral stems at the neck-shoulder junction. J. Arthroplast. **16**(2), 236–238 (2001)
41. Woolson, S., Milbauer, J., Bobyn, J.D., Maloney, W.J.: Fatigue fracture of a forged cobalt-chromium-molybdenum femoral component inserted with cement: a report of ten cases. J. Bone Jt. Surg. Am. **79**(12), 1842–1848 (1997)
42. Lam, L.O., Stoffel, K., Kop, A., Swarts, E.: Catastrophic failure of 4 cobalt-alloy Omnifit hip arthroplasty femoral components. Acta Orthop. **79**(1), 18–21 (2008)

Measurement of the Stress State in the Lower Link of the Three-Point Hitch Mechanism

Vera Cerović, Dragan Milković(✉), Aleksandar Grbović, Saša Radulović, and Jovan Tanasković

Faculty of Mechanical Engineering, University of Belgrade, 11120 Belgrade, Serbia
verica.cerovic@gmail.com, dmilkovic@mas.bg.ac.rs

Abstract. Agricultural machines and their implements are subjected to dynamic loads during farm operations. Depending on the type of operation (e.g. lifting or plowing), lower links of the three-point hitch mechanism are exposed to stresses caused by combination of bending moments and axial forces. In this paper we analyzed influence of the soil resistance during plowing in the lower link and the possibility of its failure. The stresses were measured using strain gauges at locations with uniform stress distribution in order to enable more reliable comparison with finite element analysis (FEA). Recorded stresses vs. time were used for identifying mean stresses and amplitudes for different plowing depth and different tractor speeds. Due to the geometry of the lower links and their joints in the three-point hitch mechanism, during plowing and transferring soil resistance, links are loaded not only by axial forces but also by bending moment in the horizontal plane. Under some assumptions, FEA provided us to make relations between the measured stresses and the loads that caused them. Measured stresses show that links have significant safety margin relative to tractor installed power and soil resistance, which enables the possibility of their design optimization. Obtained results may also serve for further analyses of fatigue life prediction, measurement of the draft forces etc.

Keywords: Three-point hitch mechanism · Lower link · Stress state Strain gauges · Finite element analysis

1 Introduction

Three-point hitch mechanism (TPH) is the most widely used way of connecting implements to the tractor (Fig. 1). Considering that it is extremely dynamically loaded, a substantial number of authors analyzed the possibilities of its improvement [1–3]. In this paper, the focus was on the lower link of TPH mechanism, and the first step of the analysis was to identify the stresses and the character of the loads that these elements transmit during plowing operation. Depending on the depth of plowing and the type of soil, forces needed to carry out this operation differ in a wide range. In this paper, we dealt with the first category link for low power tractors. Tractors with a power of less

N. Mitrovic et al. (Eds.): CNNTech 2018, LNNS 54, pp. 112–121, 2019.
https://doi.org/10.1007/978-3-319-99620-2_9

than 40 kW are widely used on smaller farms, and they have a significant share in the agricultural machinery market, with the tendency of increasing the number of units sold [4]. The largest investments, based on the Agrievolution Alliance data, in the past few years in this area are in the countries such as China and India.

Fig. 1. Three-point hitch mechanism (TPH) [5].

The analysis presented in this paper is based on the experimental measurements of the stresses using strain gauges during plowing operation. These values are compared to the results of finite element analysis (FEA) performed in order to determine the load intensity that caused these stresses. The measurements were made with one tractor-implement configuration, on the one soil type. Plowing is one of the most power demanding soil tillage operations. It is cutting and turning up and over the soil. Due to the small installed power of the tractor, the speed of movement was mostly limited by the depth of plowing and the resistance that tractor can overcome.

Fig. 2. Lower link of the TPH mechanism.

The tested link belongs to the first category of links, made by forging. It has holes for the pins and spherical joints at both ends. Lateral distance between the joints of the two links forming a pair on the implement side is greater than on the side closer to the tractor (Fig. 1). As a result, ends of the link, in the zone of the joints, are folded for about 6° relative to longitudinal axis of the lower link central part (Fig. 2). There are also kinematic methods for determining positions of all TPH elements [6].

2 Field Tests Preparation

Experimental measurements were performed with the IMT 539 Rakovica tractor in combination with the OLT OSIJEK PTO 2/25 plow (Fig. 3). The lower link of the support mechanism is equipped with strain gauges set and connected in the measuring bridges according to Fig. 4. The layout of the strain gauges was chosen to measure the stresses at locations with uniform stress distribution in order to enable more reliable comparison with FEA.

Fig. 3. Tractor-plow configuration.

For measurement of the stresses, strain gauges of type HBM KY31 120-3 were used (Fig. 4). Bending due to the eccentric effect of the axial force of the lower link is a consequence of link position when mounted on the tractor (Figs. 1 and 3). Due to the existence of a stabilizer that prevents the link from being tilted to the side, in addition to the axial force, there is also a force in the stabilizer, which additionally bends the lower link. Therefore, the bending moments are not only due to the geometry of the link and TPH mechanism, but also by the force in the stabilizer. One possible way of separating these influences is using blind signal separation (BSS) for recovering two or more sources that caused mixed recorded stress signal [7] or by defining analytical equations for establishing relations between the source forces and resulting stresses.

FEA confirmed that the stress state resulted from bending only due to geometry of TPH, differs in two measuring points. On the other side, considering the position of the strain gauges (the same position on the inner and outer side of the link), the relative deformation i.e. stresses caused by stabilizer are equal at both measuring points. One strain gauge is loaded by tensile and the other by the compressive strains. Each strain gauge is temperature compensated. They are connected with dummy gauge to form a Wheatstone's half-bridge. Installed link is shown in Fig. 5.

Fig. 4. Strain gauges position

Figure 6 shows complete measuring chain with used digital acquisition system (DAQ) type and laptop for data storing and visualization. Stresses were recorded at 600 Hz sampling rata and low pass filter 100 Hz was applied in order to eliminate high frequencies with low energy capability and to have clearer recorded signal.

Fig. 5. Lower link with applied strain gauges on the tractor

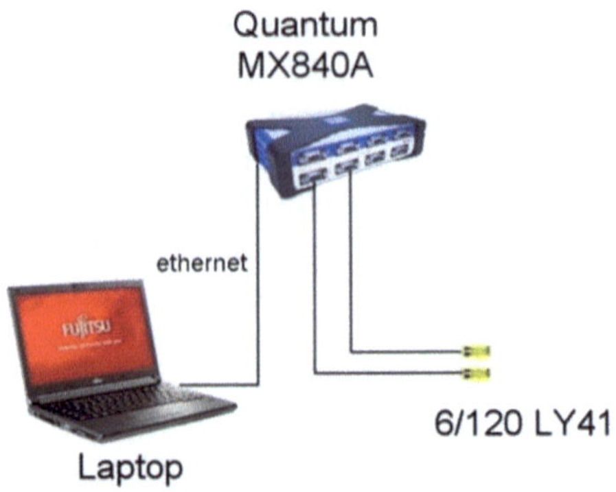

Fig. 6. Measuring chain

3 Measurement Results

Measurements in real conditions were performed on the soil of the category "chernozem" on the stubble after harvest of wheat. The test included measuring the stresses with different tractor speeds and different depths of the plowing.

The speed of the tractor was varied according to the available power in the range of 2 to 6 km/h, which could be achieved with the depth of plowing from 11 cm to 17 cm. (Table 1). Figures 7, 8 and 9 show recorded stresses vs. time for different regimes of plowing. Due to the change in the resistance of the plow, the stresses vary in a wide range. For the analysis and comparison, the measured mean stresses and amplitudes for the different depths of the plowing and the speed of the tractor's movement have been identified. Table 2 shows these values at relatively steady state. These data may serve for fatigue life analysis of the lower link using the stress-life method [3] or FEM based crack propagation method proposed in [8] and grounded on the stress intensity factors (SIFs) calculations. In the event of multiple damages, SIFs can be calculated either using extended finite element method (XFEM) or approximate method based on principle of superposition [9]. Therefore, stresses measured at lower link are necessary for fatigue life estimation as described in [3, 8, 9].

Table 1. Plowing regimes

Measurement no.	Tractor speed (km/h)	Plow depth (cm)
I	2	11
II	6	17
III	2	17

It can be seen that the stresses measured from the outside of the lower link and the inner side differ. This different comes from the link geometry and the load distribution. Therefore, the links are not quite axially loaded. There are also stresses caused by bending. The link forces are transmitted along the axis which connects the spherical joints at both ends of the links, and as a result, an additional bending of the link about approximately vertical axis occurs. Also, as explained, there exists an influence of the stabilizing chain, that prevents lateral movement during plowing. This bending appears on the span between sphere joint on the implement side and the stabilizer, as shown in Fig. 11.

Based on FEA, a relation between the stresses and forces in the lower link is established. The stress gradient in the measurement zone is small, so the possible inaccuracy of the strain gauges location does not affect the accuracy of the determination of the relative strains or the stresses at the measuring points.

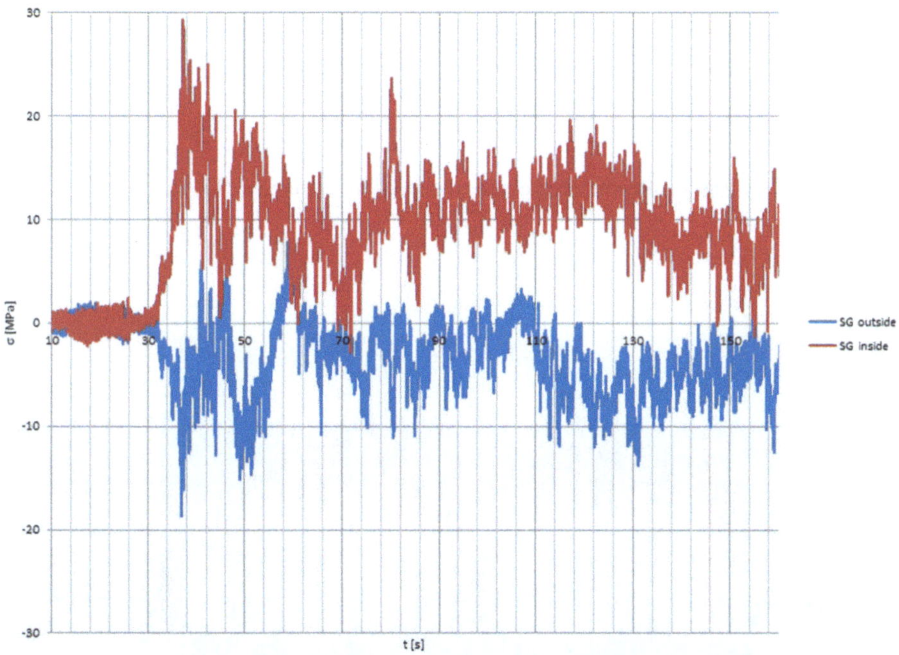

Fig. 7. Recorded stresses vs. time – measurement No. I

Figure 10 shows the finite element model for calculations and the results corresponding to measurement No. I. The maximum stress equals to 32.2 N/mm^2 appeared in the area near the sphere joint and the second critical zone is welded joint with stress equal to 25.7 N/mm^2.

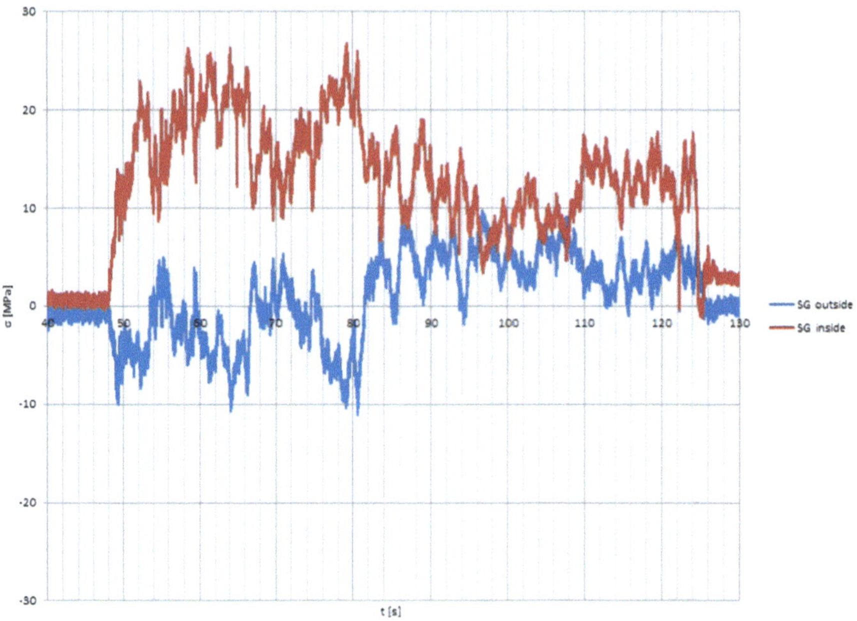

Fig. 8. Recorded stresses vs. time – measurement No. II

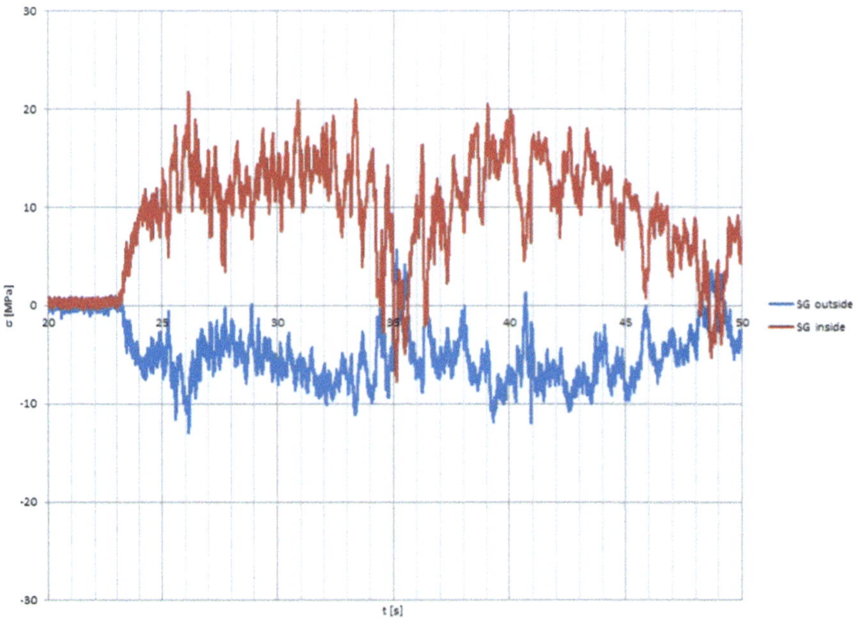

Fig. 9. Recorded stresses vs. time – measurement No. III

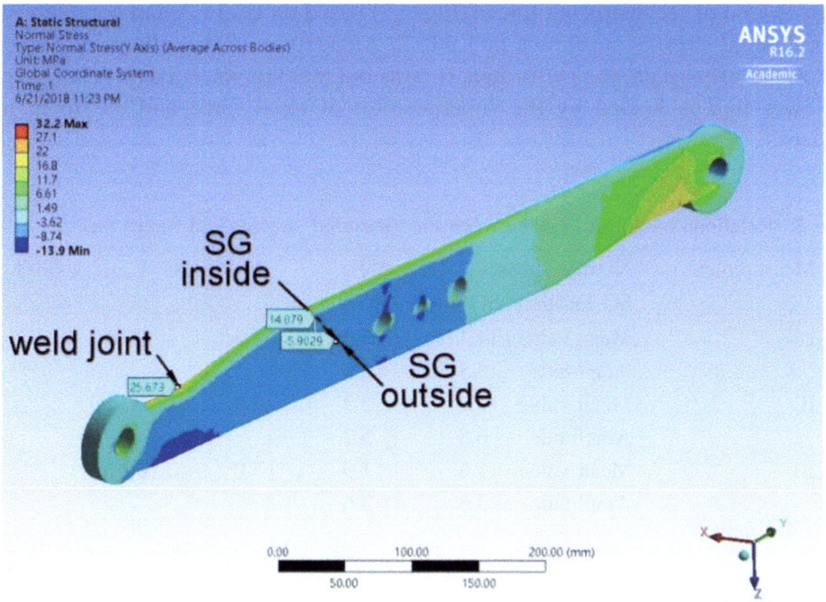

Fig. 10. FEA and obtained normal stresses during measurement No. I at the strain gauges position and at welded joint

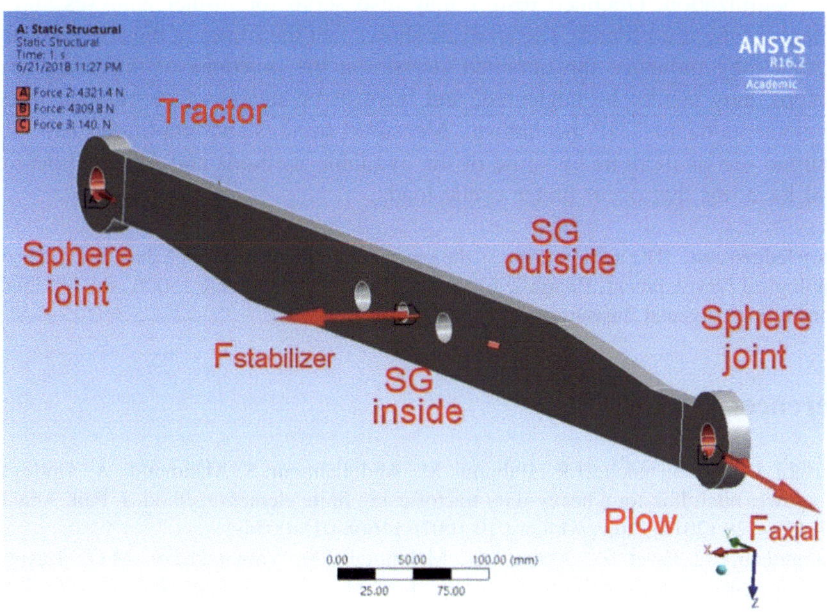

Fig. 11. Reconstructed forces based on the calculated stresses – measurement No. I

Recovered or reconstructed forces (Fig. 11) based on the FEA and relatively steady state measured stresses are shown in Table 2. An analysis of the influence of stabilizer force on the stress state was performed. It turns out that this force may not be neglected. The lower link is loaded by the superposition of axial force and the force in the stabilizer.

Table 2. Relations based on FEA between the measured stresses and forces that caused them

Measurement no.	Normal stresses σ [N/mm^2]			F_{axial} (kN)	$F_{stabilizer}$ (kN)
	SG location	SG inside	SG outside		
I	Mean value	14.1	−5.9	4.3	0.14
	Amplitude	5.5	2.5		
II	Mean value	19.9	−5.2	7.5	0.15
	Amplitude	6.5	5.2		
III	Mean value	11.6	−1.9	5.0	0.07
	Amplitude	3.8	3.6		

4 Conclusion

Measured stresses show that links are extremely variable loaded. The stresses depend on the resistance caused by tractor speed, plowing depth and the type of soil. Nevertheless, there is still significant safety margin, which enables the possibility of their design optimization. Obtained results may also serve for further analyses and measurements of the draft forces. The analysis shows that the forces in the stabilizer are not large, but they influence the obtained stresses at the positions of the strain gauges. Therefore, they cannot be neglected, and have to be measured along with forces in other elements of the THP mechanism. Measured stresses and amplitudes can be used for fatigue life predictions by some of the available methods and for identification of critical locations that occur under cyclic loads.

Acknowledgement. The authors express the gratitude to the Ministry of Education, Science and Technological Development, Republic of Serbia, Research Grants TR 35006 and TR 35045 for supporting experimental measurements.

References

1. Tarighi, J., Ghasemzadeh, H.R., Bahrami, M., Abdollahpour, S., Mahmoudi, A.: Optimization of a lower hitch link for a heavy-duty tractor using finite element method. J. Fail. Anal. Prev. **16**, 123–128 (2016). https://doi.org/10.1007/s11668-015-0054-1
2. Alimardani, R., Fazel, Z., Akram, A., Mahmoudi, A., Varnamkhasti, M.G.: Design and development of a three-point hitch dynamometer. J. Agric. Technol. **4**(1), 37–52 (2008)
3. Dalmiş, I.S., Tezcan, O., Eruslu, S.Ö.: Fatigue life enhancement of three point hitch system brackets in the garden series tractors. J. Agric. Sci. **23**, 185–194 (2017)

4. Lowder, S.K., Skoet, J., Raney, T.: The number, size, and distribution of farms, smallholder farms, and family farms worldwide. World Dev. **87**, 16–29 (2016). https://doi.org/10.1016/j.worlddev.2015.10.041
5. John Deere. https://www.deere.com/en/. Accessed 22 June 2018
6. Ambike, S.S., Schmiedeler, J.P.: Application of geometric constraint programming to the kinematic design of three-point hitches. Appl. Eng. Agric. **23**(1), 13–21 (2007)
7. Milkovic, D., Simic, G., Jakovljevic, Z., Tanaskovic, J., Lucanin, V.: Wayside systems for wheel-rail contact forces measurements. Measurement **46**, 3308–3318 (2013). https://doi.org/10.1016/j.measurement.2013.06.017
8. Grbovic, A., Rasuo, B.: FEM based fatigue crack growth prediction for spar of light aircraft under variable amplitude loading. Eng. Fail. Anal. **26**, 50–64 (2012). https://doi.org/10.1016/j.engfailanal.2012.07.003
9. Kastratovic, G., Grbovic, A., Vidanovic, N.: Approximate method for stress intensity factors determination in case of multiple site damage. Appl. Math. Model. **39**(19), 6050–6059 (2015). https://doi.org/10.1016/j.apm.2015.01.050

Engineering

Research of Lean Premixed Flame by Chemiluminescence Tomography

Vuk Adžić, Mustafa Makhzoum, Aleksandar Milivojević(✉), and Miroljub Adžić

Faculty of the Mechanical Engineering, Fuel and Combustion Laboratory, University of Belgrade, 11120 Belgrade, Serbia
{vadzic, amilivojevic}@mas.bg.ac.rs

Abstract. An experimental research of lean premixed flame in a swirl burner is performed using CH* chemiluminescence imaging. The images of flame are acquired and the flame structure is investigated. The position of flame front, represented by the area with highest rate of exothermic chemical reactions, is deduced from the acquired images. To identify the flame front using two dimensional images, a method of image tomography is applied. A tomography of the images is numerically performed using an in-house developed numerical Abel transform. It is found that flame tomography enables precise location of flame front in turbulent flows. The tomography research is compared and analyzed with the numerical results of the same burner.

Keywords: Premixed flame · Swirl burner · Chemiluminescence Tomography · Abel transform

1 Introduction

In order to control emissions of oxides of nitrogen and carbon monoxide, one of the proven methods is to prepare a lean mixture of fuel and air and burn in furnaces, boilers, gas turbines or household appliances. Lean combustion decreases flame temperature and laminar flame velocity and can induce problems of flame stabilization. There are different methods to stabilize a flame, such as, step and bluff body flame holders, auxiliary flame and swirling flow. The swirling flow method has proved very good performance but lean combustion still can induce instabilities which are prohibitive in practical combustion systems. It is therefore very important to analyze the flame structure, appearance of local flame extinction and behavior under influence of variables such as fuel composition, air coefficient, and thermal power, fuel/air mixture velocity field, pressure, preheat temperature, burner geometry [1–6]. The goal of this paper is to present the optical method for flame investigation and the experimental research results of a lean premixed flame in a swirl burner, aiming at precise flame location.

The paper deals with experimental and computational study of swirling lean premixed flame. The research uses CH* imaging to reveal flame geometry and location and to compare experimental findings with the numerical results research. The CH* chemiluminescence imaging is a nonintrusive optical method that enables insight in flame details with good resolution capabilities [1]. In swirling flames, the acquisition of

N. Mitrovic et al. (Eds.): CNNTech 2018, LNNS 54, pp. 125–136, 2019.
https://doi.org/10.1007/978-3-319-99620-2_10

CH^* chemiluminescence is the line-of-sight integration of local CH^* emissions. In order to calculate the value of local volumetric CH^* emission it is necessary to reconstruct the obtained image using the Abel transform. The imaging method, part by part of an object, i.e. flame, is known as a tomography [7–10]. Development of digital cameras with CCD sensors brought new possibilities for optical research of flames. Nevertheless, CCD sensors and digital cameras have their limitations which are also limitations of their use in flame research [10]. A CCD sensor and a camera are optically sensitive only in certain limits of wavelengths. There are sensors and cameras purposely designed for infrared or ultraviolet light spectra. Standard CCD sensors and cameras are sensitive in the visible light spectra. In combination with optical filters a standard camera can be used for more specific flame research. Beside flame geometry, the CH^* chemiluminescence imaging can identify zone of highest heat release rate, local air coefficient, temperature and relative concentration of some chemical species [4–6, 10].

2 Flame Imaging Method

The light emission in ultraviolet and visible light spectra, due to chemical reactions, is known as chemiluminescence. It is caused by electronically excited short living chemical species such as OH^* (310 nm), CH^* (388 and 430 nm) and C_2^* (473 nm), Fig. 1. In this research CH^* is used as a marker, because it its good performance in lean mixtures [8].

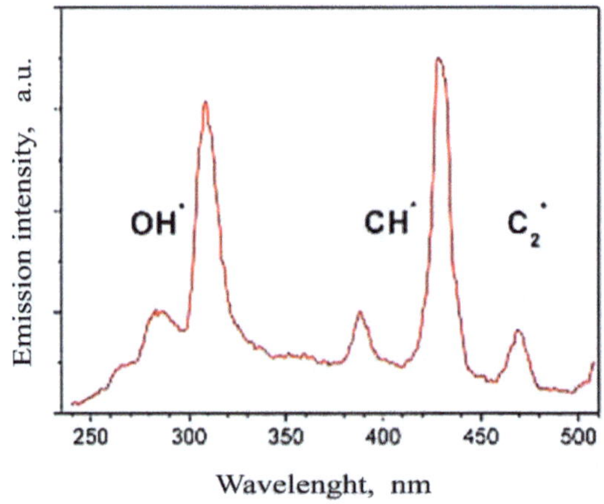

Fig. 1. Chemiluminescence species: OH^* (310 nm), CH^* (388 and 430 nm), C_2^* (473 nm) [1].

The main reaction path of CH^* production is

$$C_2H + O = CH* + CO$$

The complete model of CH* production is still not fully developed, contrary to the CH* destruction reactions model in which CH radical is the main product. Therefore, CH radical may be used as the indicator of CH* presence. The experimental findings confirm that the maximum concentration of

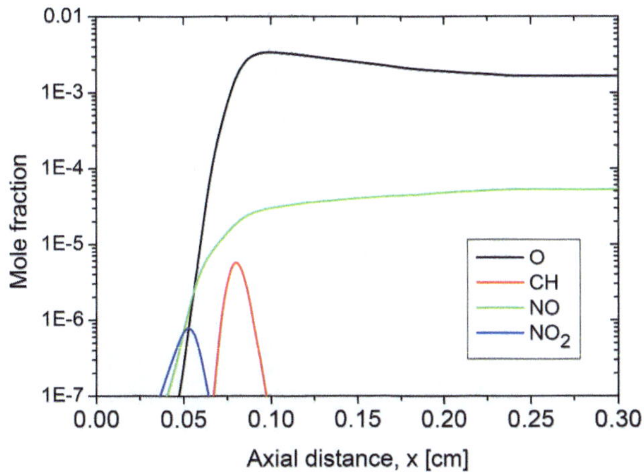

Fig. 2. Selected species profiles in premixed methane flame front, air coefficient 1.0.

CH* is in the flame zone of high heat release rates, therefore CH* can be used as the flame location indicator [1, 4, 6, 11]. Figures 2 and 3 [7] show calculated (CHEMKIN) premixed methane-air flame front and the selected chemical species profiles.

It can be seen, Fig. 2, that the CH radical is present in the 0.7–1.0 mm zone. In the same zone, Fig. 3, CH_4 is fully consumed and O_2, H_2O, CO and CO_2 reach their equilibrium concentrations. This confirms the assumption that CH* can be used as the flame location indicator.

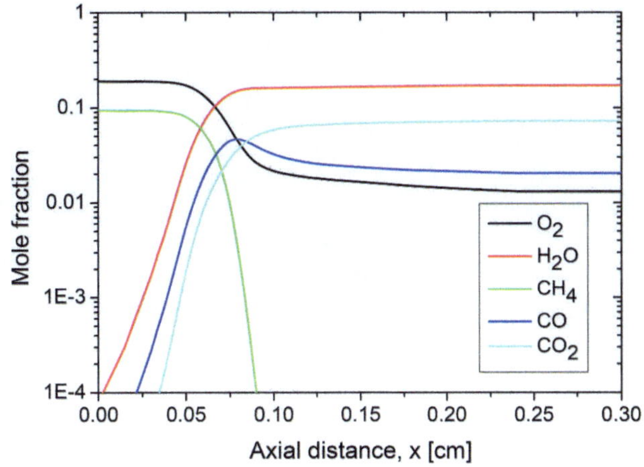

Fig. 3. Selected species profiles in premixed methane flame front, air coefficient 1.0.

Let the intensity of CH* chemiluminescence is

$$i = i(x, y, z)$$

The chosen coordinate system and position of the CCD sensor is presented in Fig. 4.

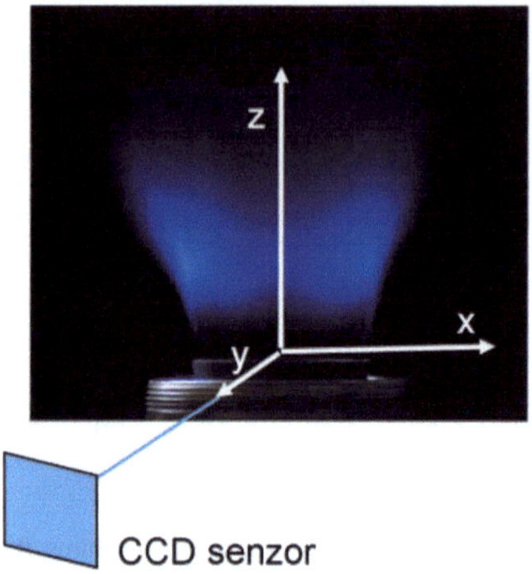

Fig. 4. The flame coordinate system and CCD sensor.

We assume that the flame is ax symmetric for any cross sections, and therefore, the local intensity of chemiluminescence i is function of x, r, $i\ (x,r)$. The value of $I(x,r)$, acquired by CCD sensor, related to the line-of-sight s, Fig. 5, is

$$I(x,r) = \int_0^R i(x,r)\frac{2r}{(r^2 - x^2)^{0.5}}dr \tag{1}$$

$$r^2 = x^2 + y^2 \tag{2}$$

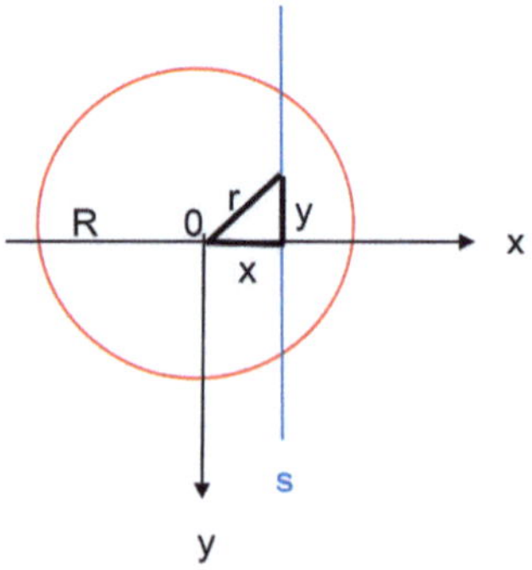

Fig. 5. Line-of-sight, s, and the coordinate system.

which is actually the CH* emission acquired by a CCD sensor. As we are interested in local values of intensity of light emission per unit of volume, $i = i\ (x,r)$, we need the inverse value of $I(x, r)$. The analytical inverse

$$i(r) = -\frac{1}{\pi}\int_0^R \frac{\frac{dI(r)}{dr}}{(r^2 - x^2)^{0.5}} dr \tag{3}$$

is known as the Abel transform. The Abel transform has no straightforward solution. There are different approaches in solving this problem: spline, polynomial, fast Fourier transform, numerical and still new ideas appear. The authors of this paper have developed and applied a numerical method. The method is based on the assumption of ax symmetric flame and therefore ax symmetric distribution of CH^* emission, $i = i(r)$. The numerical approach is as follows: The flame is divided into concentric rings which thicknesses are Δx. Starting from the first, outer ring, Fig. 6, which projected area is calculated and light integrated intensity I_1 acquired, one gets the value of $i = i\ (r_o)$. Using the same procedure for the rings towards r = 0, with calculated areas and $i = i\ (r)$ emissions of all previous rings, one can calculate the unknown value of CH^* emission, using the same procedure until the last ring.

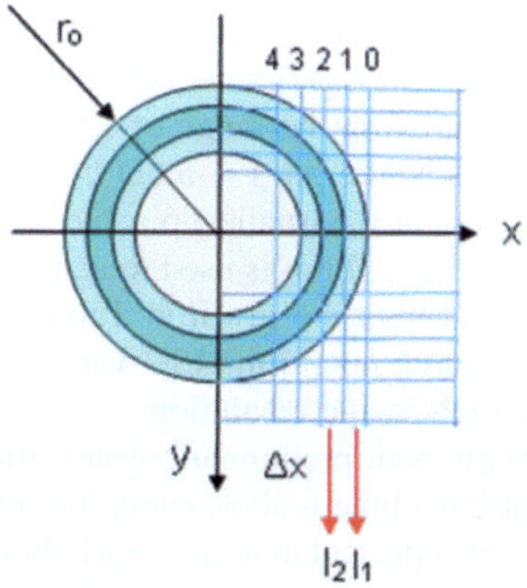

Fig. 6. Schematic of flame.

The simulated results of this Abel transform has an error of the order ±1%.

3 Experimental

For the experimental research, a purposely built swirl burner is used, schematically given in Fig. 7. The fuel and air premixing is completed in a mixing part of burner and then injected into free space through the axial swirler (vane angle of 45° and swirl number 0.7). The burner is fueled with propane-air or methane-air mixtures with controlled flow rates, air coefficient and thermal power. The propane/methane quality is 99.5% by volume.

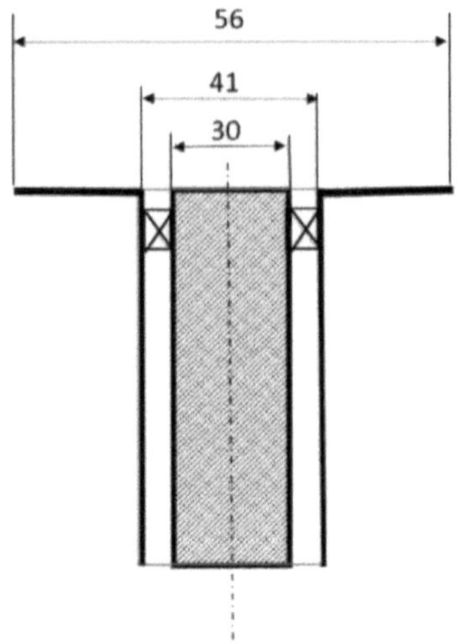

Fig. 7. Schematic of the swirl burner.

The experimental setup consists of fuel cylinders, air fan, fuel and air flow rates measuring and control systems, temperature and pressure measuring systems. The error of measurements is for the volume flow rates of fuel ±5%, air ±5%, temperature ±0.5 °C and pressure ±10 Pa. The maximum nominal thermal power of burner is 12 kW. The air coefficient is varied in the range of 1.0–1.6.

A Nikon D80 camera with a NikkorAF f 4–5.6D ED 70–300 mm telephoto lenses is used for CH* chemiluminescence visualization and acquisition. An optical filter Ealing 35–3300 centered at 430 ± 2.0 nm is used to acquire CH* emission. Technical details of camera, telephoto and filter are given in Tables 1, 2, and 3. The camera is very suitable for the flame research due to the ISO variability, span of shutter speeds, electronic amplification and high image resolution.

The image acquisition starts with preliminary systematic variation of aperture and shutter speed. It is found that 2 ms time is short enough to get the turbulent structure of flame, while 3 s for an average representative image of about 1000 images. In order to compare emissions, preliminary shots are used to adjust ISO sensitivity and shutter speed in such a way that the maximum value of the acquired CH^* light value is less than 250 on an 8-bit scale, (0–255). The effective image format is 249 pixels per line (x – coordinate) and 204 pixels per field (z – coordinate). This corresponds to 0.47 mm/pixel.

Table 1. Camera technical data

Camera DSLR	Nikon D80
Sensor	APS-C 10.8MP/10.2MP eff, 23.6 mm × 15.8 mm Nikon DX format
Resolution	3.872 × 2.592 (10.2 effective megapixels)
ISO	100–1600
Shutter speed	30 to 1/4000 s
Frame acq. rate	3 1/s, 100 JPEG/6 RAW

Table 2. Telephoto lenses technical data

Telephoto	NikkorAF f 4-5.6D ED
Zoom	70–300 mm
Max. aperture	f/4–5.6
Min. aperture	f/32
Min. distance	1.5 m
View angle	34.20–8.10°

Table 3. Filter technical data

Optical filter	Ealing 35–3300
Wavelength, nm	430.0 ± 2.0
Filter diameter, mm	25.4
Transparency	51%
Attenuation of other wave lengths	>99%

4 Results and Discussion

Figure 8 shows typical flame images of swirling premixed flame. The swirling flame is characterized by an inner recirculation zone which stabilizes flame. It can be seen that the zone with highest rates of exothermic chemical reactions is conical.

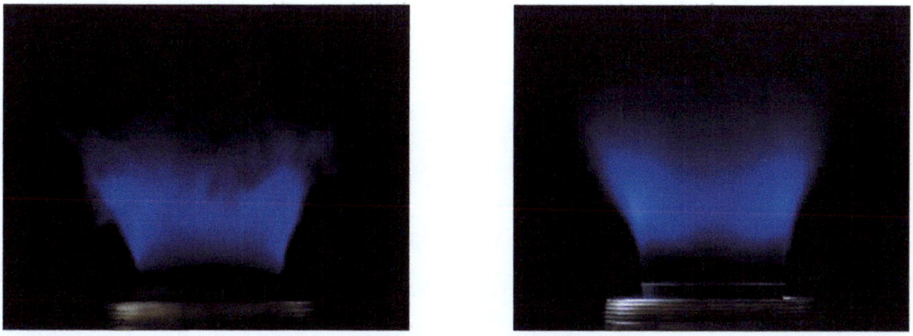

Fig. 8. CH* chemiluminescence images. Propane–air, 4 kW thermal power, air coefficient 1.0. Shutter speeds: 2 ms, left image and 3 s, right image

The left image shows a turbulent flame, while the right image is an averaged image of the order of 10^3 instant images. The intensity of acquired CH* chemiluminescence emission is presented on 0–255 scale using a raw image data and an in-house developed software, as shown in Fig. 9. A horizontal x line containing a pixel with highest value of emission intensity is chosen as the

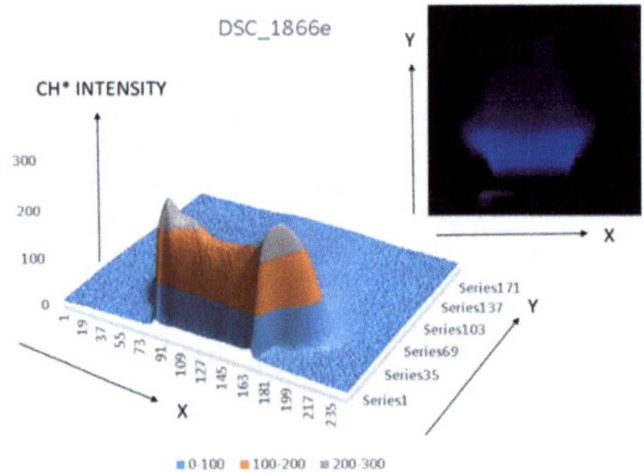

Fig. 9. The coordinate system and CH* chemiluminescence emission distribution. Propane–air, 4 kW thermal power, air coefficient 1.0, [10].

Indicator of light intensity for the whole set of tests. Then the marker intensity is amplified to the 245 value and all of acquired images are multiplied by the same value of amplification, Fig. 10. This procedure enables relative comparison of CH* chemiluminescence emissions for each set of tests.

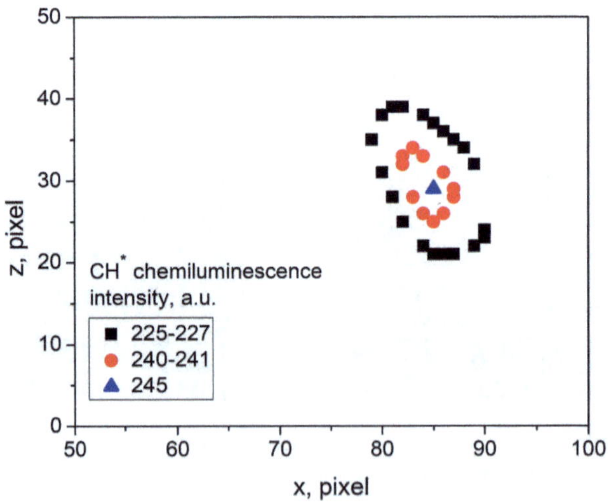

Fig. 10. Isolines of CH* chemiluminescence vs x and z coordinates. Propane-air, thermal power 4 kW, air coefficient 1.0.

Typical CH* chemiluminescence emission as a function of x, is shown in Fig. 11. The figure reveals noise of the CCD sensor as a function of CH* emission intensity. It can be seen that the noise is pronounced at low CH* emission intensity. The error due

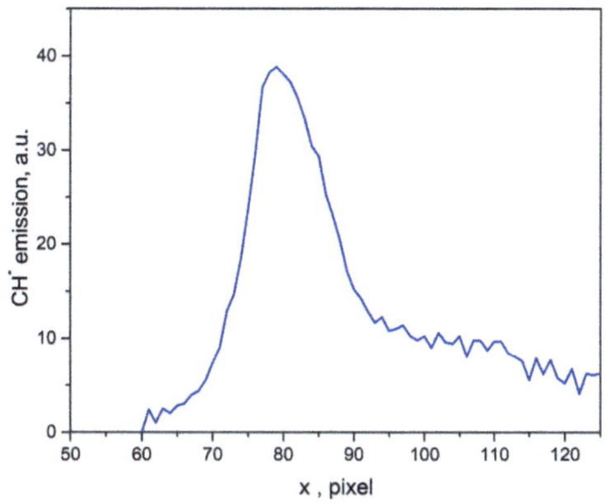

Fig. 11. CH* chemiluminescence emission as a function of x, i.e. r coordinate. Propane-air, thermal power 4 kW, air coefficient 1.0.

to the sensor noise is ±3% [10]. As an example, the case of propane-air premixed flame (z = 38), 4 kW thermal power, air coefficient 1.0 acquired CH* emission in arbitrary units, and the Abel transform, which values are multiplied by 10 for better visibility, is shown in Fig. 12. It is interesting to compare the experimental and numerical results for the same burner and same operating parameters. Figure 13 shows that the numerically obtained maximum temperature location coincides with the maximum value of CH* emission. Besides, the applied CFD ANSYS software uses reduced chemistry in order to keep the time of calculations in acceptable limits. After calculating the

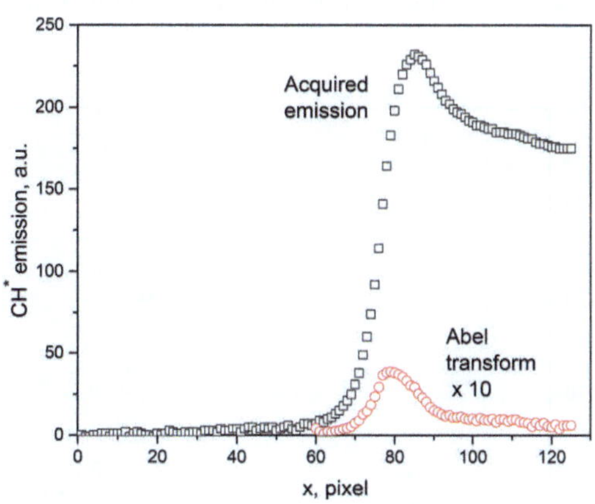

Fig. 12. The CH* acquired emission and Abel transform, propane-air premixed flame, 4 kW thermal power, air coefficient 1.0. The Abel transform values are multiplied by 10 for better visibility.

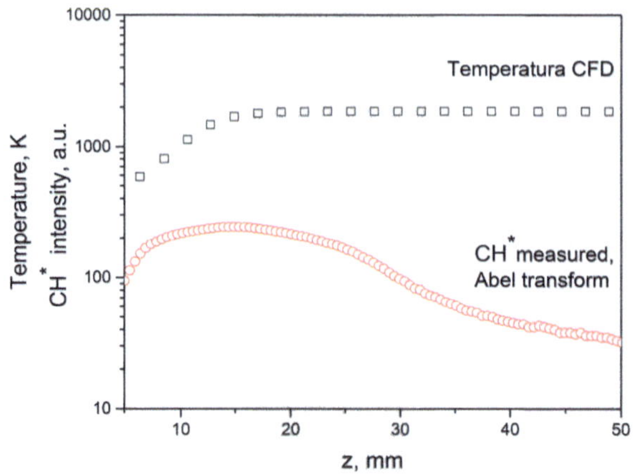

Fig. 13. CFD temperature and CH* emission tomography comparison vs. z coordinate [11].

Maximum value of temperature, kept the temperature constant for z > 18. As opposed to the numerical method of flame research, the CH* tomography shows that after the maximum value of CH* emission, there is a decrease which reveals the tomography capabilities of precise flame location in turbulent flows. The effect of fuel on CH* emission, at constant thermal power and air coefficient, is shown in Fig. 14. The intensity of CH* emission is higher for propane because the main reaction path to CH* is via C_2H radical. Higher carbon to hydrogen molar ratio of 0.375 for propane vs. 0.25 for methane facilitates CH* production. Figure 14 shows that the geometry and location of both flames practically do not depend on fuel used. Although laminar flame speeds of stoichiometric air mixtures of propane and methane differ, 0.43 and 0.37 m/s,

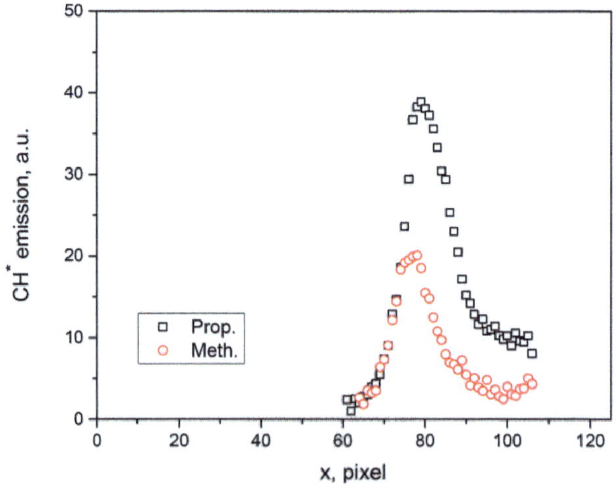

Fig. 14. CH* emission Abel transform of propane-air (at z = 38) and methane-air (at z = 33 pixels) premixed flame. 4 kW thermal power, air coefficient 1.0.

respectively, in turbulent regime, the locations of both flames do not differ, Fig. 14. This confirms that the turbulent flame speeds of hydrocarbons depend mainly on turbulence, [12–14].

5 Conclusions

The experimental research of lean premixed flame in a swirl burner, using CH* chemiluminescence tomography, has been performed. A digital camera and an optical filter centered at 430 nm are used for flame imaging. An Abel transform numerical method is developed and applied for image processing. Propane and methane were used as typical representatives of hydrocarbon gaseous fuels.

The experimental results show that the CH* tomography is a valuable tool for flame research. It enables precise identification of heat release zones, flame location and flame geometry.

The intensity of CH* emission is higher for propane due to the main reaction path to CH* production via C_2H radical and more favorable carbon to hydrogen molar ratio of propane.

The geometry and location of flames practically do not depend on fuel used. This confirms that the turbulent flame speeds of hydrocarbons depend mainly on turbulence.

When compared with the numerical method of flame research and analysis CH* tomography shows superior capabilities for precise flame location in turbulent flows.

References

1. Gaydon, A., Wolfhard, H.: Flames, Their Structure, Radiation and Temperature. Chapman and Hall, London (1960)
2. Gupta, K., Lilley, G., Syred, N.: Swirling Flows. Abacus Press, London (1984)
3. Adzic, M., Fotev, V., Zivkovic, M., Milivojevic, A.: Effect of a microturbine combustor type on emissions at lean-premixed conditions. J. Propuls. Power (2010). https://doi.org/10.2514/1.47456
4. Taamallah, S., Chakroun, N., Watanabe, H., Shanbhogue, S., Ghoniem, A.: On the characteristic flow and flame times for scaling oxy and air flame stabilization modes in premixed swirl combustion. Proc. Combust. Inst. **36**(3), 3799–3807 (2016). https://doi.org/10.1016/j.proci.2016.07.022
5. Carlsson, H., Nordström, E., Bohlin, A., Wu, Y., Zhou, B., Li, Z., Alden, M., Bengtsson, P. E., Bai, X.-S.: Numerical and experimental study of flame propagation and quenching of lean premixed turbulent low swirl flames at different Reynolds numbers. Combust. Flame **162**, 2582–2591 (2015). http://lup.lub.lu.se/record/8080551
6. Hardalupas, Y., Orain, M.: Local measurements of the time-dependent heat release rate and equivalence ratio using chemiluminescent emission from a flame. Combust. Flame **139**(3), 188–207 (2004)
7. Adzic, M., Fotev, V., Jovanović, S., Milivojević, A., Jovicic, V., Milekic, G., Bogner, M., Adzic, V.: The FP6-INCO-WBC Flexible Premixed Burners for Low Cost Domestic Heating Systems. FLEXHEAT, Supported by the EC (2004–2007)
8. Orain, M., Hardalupas, Y.: Effect of fuel type on equivalence ratio measurements using chemiluminescence in premixed flames. Comptes Rendus Mec. **338**(5), 241–254 (2010)

9. Guibert, T., et al.: Flame chemiluminescence from CO_2—and N_2-diluted laminar CH_4/air premixed flames. Combust. Flame **181**, 110–122 (2017)
10. Adzic, V., Milivojevic, A., Adzic, M.: Flame diagnostics making use of visualization of CH group. In: Proceedings of IEEP 2017 Conference, Zlatibor, June 2017 (2017)
11. Makhzoum, M., Adzic, M., Fotev, V., Milivojević, A., Adzic, V.: Numerical analysis of lean premixed combustor fueled by propane-hydrogen mixture. Thermal Sci. (2017). https://doi.org/10.2298/TSCI160717131M
12. Adzic, M., Carvalho, I., Heitor, M.: Visualisation of the disintegration of an annular liquid sheet in a coaxial airblast injector at low atomizing air velocities. Opt. Diagn. Eng. **5**(1), 27–38 (2001)
13. Adzic, M., Carvalho, I., Heitor, M.: Error analysis and calibration procedure when using an ICCD camera for the study of spray formation. J. Flow Vis. Image Process. **4**, 149–162 (1997)
14. Adzic, M., Zivkovic, M., Fotev, V., Milivojevic, A., Adzic, V.: Influential parameters of nitrogen oxides emissions for microturbine swirl burner with pilot burner. Chem. Ind. **64**(4), 357–363 (2010)

freeCappuccino - An Open Source Software Library for Computational Continuum Mechanics

Nikola Mirkov[1(✉)], Nenad Vidanović[2], and Gordana Kastratović[2]

[1] Institute of Nuclear Sciences – Vinča, Laboratory for Thermal Engineering and Energy, University of Belgrade, 11351 Belgrade, Serbia
nmirkov@vin.bg.ac.rs
[2] Faculty of Traffic Engineering, University of Belgrade, 11000 Belgrade, Serbia

Abstract. The paper describes development of an open-source library (www.github.com/nikola-m/freeCappuccino) for computational fluid dynamics and in general computational continuum mechanics. The code is based on finite volume method on arbitrary unstructured polyhedral meshes. The interfaces to highly abstract data types such as arbitrary order tensor fields on discretized finite volume domains, and scalar and vector sparse linear systems resulting from finite volume discretization of partial differential equations are provided. Explicit manipulation of tensor fields through high level, highly abstract programming syntax is explained. Also, implicit operation over tensor fields pertinent to discretization of partial differential operators is provided and explained. The library is developed in modern version of Fortran. Code parallelization is achieved through domain decomposition and implemented using MPI and OpenMP. While avoiding the usual class syntax of object-oriented programming, the code has essentially object oriented design. Comparison is made with the well-known OpenFOAM library. The purpose of the ongoing development is providing researchers with a tool for easy transfer of mathematical operations of their physical models into functional and efficient simulation software based on finite volume method. The guiding principle of development is exchange of ideas and reproducibility in computational science in general.

Keywords: Engineering software · Computational fluid dynamics
Finite volume method · Parallel computing · High-performance computing

1 Introduction

We are in the age when engineering and science rely to a large extent on computational tools. Many of these tools are developed within a single commercially oriented company and source code stays out of the reach of engineers and scientists. In efforts to simulate more complex physics researchers are often able to add additional functionality trough appropriate interface, while in most such cases, the core simulation code is inaccessible to the user. Influential voices are being heard recently [1] in scientific community, where the important question of reproducibility in science is related, when it comes to computational science, to accessibility of the source code of the program

N. Mitrovic et al. (Eds.): CNNTech 2018, LNNS 54, pp. 137–147, 2019.
https://doi.org/10.1007/978-3-319-99620-2_11

which was used to generate the simulation data. It is stressed there that the open-source software libraries constitute the backbone of reproducible scientific computing, and that not releasing original source code raises needless roadblocks to reproducibility.

Our effort in creating a platform for collaborative development of computational algorithms and testing of mathematical models results in freeCappuccino library. The name is derived to assert two sources that influenced its development, and also to suggest that we are dealing with CAFFA (Computationally Aided Fluid Flow Analysis) [2] with FOAM (Field Operation and Manipulation) [3, 4].

While the mature commercial software tools enable efficient treatment of multi-disciplinary design optimisation (e.g. [5]) due to its monolithic character, open-source tools tend to use distributed effort on modular components written by field experts on a common platform. The challenges faced when developing open-source software tools are widely discussed [6].

Solution of complex problems of continuum mechanics (fluid flow, heat transfer, elasticity, aeroacoustics) and electromagnetics relies on approximate solution of governing Partial Differential Equations. A successful library in this field should be based on a flexible numerical approach, to be able to handle complex geometric domains found in practical situations, to have data structures capable of representing basic objects of study in the studied discipline (in our case tensor fields), and to be able to scale efficiently on HPC hardware. Also it is important to easily interface other tools in the chain, whether pre-, post-processing software or other computational libraries.

In the case of freeCappuccino we have decided to base our approach on Finite Volume Method for approximation of governing equations and to use essentially object oriented approach. Other design decisions followed from these basic ones. We have also decided to write code in modern Fortran, since it enabled implementation of all the ideas that emerged along the way and has good performance record.

2 Handling Complex Geometry and Computational Meshes

The requirement of ability to treat complex geometries which abound in engineering applications led to various numerical approaches - immersed material boundaries, castellated, and body fitted meshes. Body fitted meshes evolved to the concept of generally unstructured polyhedral meshes as the culmination of its development. In this approach, continuous domains are subdivided into discrete volume elements of polyhedral shape with arbitrary number of faces. In such case it is useful to have so-called 'face-based' data structure (not the only option for body-fitted meshes cf. edge-based data structure in mean-median, and Voronoi dual meshes). The face-based data structure is useful also in other approaches where, for example, we have hexahedral cells and adaptive mesh refinement based on oct-tree cell division, or in cases where block-wise structured meshes are interfaced in non-conformal way (cell faces at block interface don't match).

While it is relatively straightforward to compute geometric properties when computational cells are well defined objects like tetrahedra, prisms, pyramids and hexahedra, it is not a trivial task in the case of face-based data structure, where optional cases include general polyhedral meshes or meshes obtained through adaptive mesh refinement. In the present code the basic geometric data required by the Finite Volume Method, i.e. cell volumes, coordinates of cell centres and cell face centres, components of cell face normals scaled by the magnitude of the face area projection in the same coordinate direction, are all computed based on face data, with no reference to specific cell shape. For cell volume and cell centroid computation, the approach presented in [7], where, only face properties such as areas, unit normals, and face centres are employed, is adopted in present library.

Specifically important for Finite Volume Method is the way interpolation factors λ_i, i = 1, numFaces, are calculated. In the present case they represent the distance from the point where line joining cell centres intersects plane of the cell face (point j′), divided by the distance of cell centres, Fig. 1. The point of intersection is found first using the analytic expressions for the intersection of a line, and a plane, rewritten in terms of points defining the plane and the straight line. This point is not necessarily within the cell face, based on the level of cell nonorthogonality.

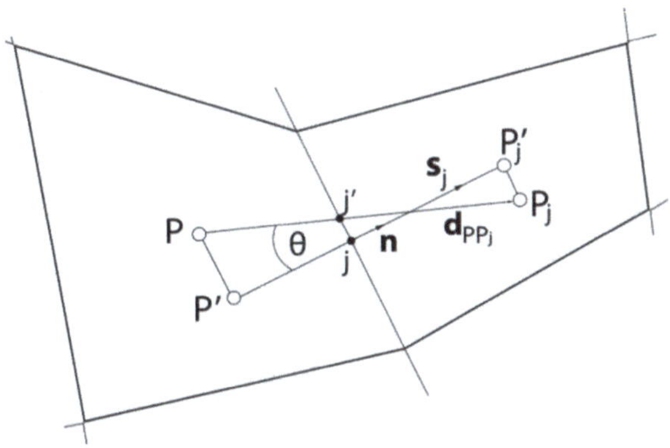

Fig. 1. A nonorthogonal grid arrangement, here represented in 2D, where the basic geometric parameters are displayed

The polyMesh format used by OpenFOAM is mesh format of choice for the library, because of its intelligent design approach which eliminates the need for cell connectivity data. The only point in simulation where cell connectivity is important is for transferring present face-based geometry data to cell-based one while writing Paraview native unstructured mesh (.vtu) files which is cell based. For that purpose cell connectivity data is generated by built in cellConectivity software utility. Other choices for input file format include Gmsh, which can be read by built-in Gmsh reader. The difference is that this format doesn't contain cell-face owner-neighbour pair data, crucial for finite volume method, and is generated before the simulation run.

It is clear that face-based approach in definition of mesh geometry gives great flexibility and requires no assumption on the type of the mesh elements, therefore such an approach is highly favourable if one insists in general application.

3 Tensor Fields Manipulation

Most important objects in the library are scalar, vector and tensor discrete fields, i.e. fields with prescribed values on discrete polyhedral cell mesh elements. There are two types of tensor fields considered, volume fields - those pertinent to cell centers, which usually denote cell averaged values in Finite Volume Method, and surface fields - where given values are defined at face centroid, or are otherwise enumerated by cell face index. In Fortran syntax they are defined using derived data types with allocatable elements. Here is the definition of three basic volume tensor field types, the third one is the symmetric tensor field type often in use in continuum mechanics (e.g. strain-rate tensor).

```
type volScalarField
  character(len=30) :: field_name
  real(dp), dimension(:), allocatable :: mag
end type

type volVectorField
  character(len=30) :: field_name
  real(dp), dimension(:), allocatable :: x, y, z
end type

type volSymmetricTensorField
  character(len=30) :: field_name
  real(dp), dimension(:), allocatable :: xx, xy, xz
  real(dp), dimension(:), allocatable ::     yy, yz
  real(dp), dimension(:), allocatable ::         zz
end type
```

A number of the basic arithmetical operators are overloaded so one can add, subtract and multiply given fields. Besides overloading, operators are polymorphic in character, so multiplication sign ‘*’ can mean scalar field - vector field multiplication, or scalar - rank two tensor multiplication, depending on type of the fields on two sides of the ‘*’ sign. Also a number of basic vector and tensor field algebra operations are defined. These are defined as binary and unary operators, see Table 1. Polymorphism is used with these operators to seamlessly treat distinct cases when different ranks of tensors are given, or different tensor type (volume, surface, symmetric, etc.) are provided.

As an example we present what a call to unary operator for computation of a deviatoric part of a rank 2 tensor field -.dev. incorporates.

Table 1. Basic tensor field operators in freeCappuccino library.

Operator	Type	Operates on
.dot.	Binary	Vol/surface rank 1/rank 2 tensor fields
.cross.	Binary	Rank 1 tensor fields
.tensor.	Binary	Vol/surface rank 1/rank 2 tensor fields
.colon.	Binary	Vol/surface rank 2 tensor fields
.transposed.	Unary	Rank 2 tensor fields
.trace.	Unary	Rank 2 tensor fields
.det.	Unary	Rank 2 tensor fields
.diagonal.	Unary	Rank 2 tensor fields
.hodge.	Unary	Rank 2 tensor fields
.curl.	Unary	Rank 2 tensor fields
.symm.	Unary	Rank 2 tensor fields
.skew.	Unary	Rank 2 tensor fields
.magSq.	Unary	Rank 2 tensor fields
.mag.	Unary	Rank 2 tensor fields
.dev.	Unary	Rank 2 tensor fields
.devTwo.	Unary	Rank 2 tensor fields
.hyd.	Unary	Rank 2 tensor fields

```
function deviatoric_part_rank2_tensor(T)  result(devT)
  implicit none
  type(volTensorField), intent(in) :: T
  type(volTensorField)             :: devT
  type(volTensorField)             :: I

  devT = new_volTensorField(numCells)
  I    = eye(numCells)
! overloaded operator - here '-' substracts two tensor fields
!          | '*' multiplies tensor field by a constant scalar
!          |              | '*' multiplies tensor fields by a scalar array
!          |              |         |
  devT = T - ( 1./3.0_dp * ( .trace.T * I) )

end function deviatoric_part_rank2_tensor
```

In the calling routine there are other examples of calls to unary and binary operators and polymorphism.

The main appeal for operator overloading and of shorthand operator definitions is to make it easy for users to remember how to call tensor algebra operations and to make the written code easier to read and interpret [8]. In this way the overloaded and shorthand operators have their conventional meaning and expressions of manipulations on tensor fields get their compact, comprehensive form.

4 Finite Volume Discretisation

The finite volume method for approximation of PDEs has established itself as a method of choice for engineering applications of computational fluid mechanics due to its flexibility, and most of the general purpose CFD codes employ this method for equation approximation.

To systematize the approximation of the governing equations of continuum mechanics, we cast them into integral form of the general conservation law, which is one of the central general objects of approximation,

$$\frac{d}{dt}\int_V \rho\phi dV + \int_S \rho\phi\mathbf{u}\cdot\mathbf{n}dS = \int_S \Gamma_\phi\nabla_\phi\cdot\mathbf{n}dS + \int_V q_\phi^V dV + \int_S \mathbf{q}_\phi^S\cdot\mathbf{n}dS \quad (1)$$

where $\boldsymbol{\varphi}$ is either a scalar, or a vector or tensor component to which conservation law is applied. The two terms on the left are the transient and convection term, on the right are diffusion term, and volume and surface source terms. Variables figuring in this equation are: fluid density $\boldsymbol{\rho}$, **t** is time, u fluid velocity vector, $\boldsymbol{\Gamma}$ is diffusion coefficient, and finally volume and surface source terms q^V and $\mathbf{q}^S$, where surface term is a vector quantity.

The development of the numerical algorithm for freeCappuccino library has been described in detail in [9], where most of the algorithmic improvements are covered and code verification and validation details are presented. Code was specifically targeted for application on highly complex nonorthogonal meshes, therefore proving the second order accuracy, even in highly distorted mesh situations, was specially considered in the paper. The verification tests ranged from manufactured solutions to analytical solutions.

5 Linear Algebra

The process of discretisation of governing PDEs leads to linear systems of equations with sparse matrices. To represent such a mathematical object, an appropriate storage format should be chosen, leading to efficient storage (memory) requirements, in correspondence to numerical method and an easy interface with external libraries. The Compressed Sparse Row (CSR) format [10] as probably most popular among general sparse matrix storage formats is chosen for freeCappuccino library. Such choice is based, in the first place, on considerations of underlying finite volume algorithm. Additional advantage is based on the fact that CSR format is used by many numerical linear algebra libraries.

The finite volume discretisation involves calculating fluxes between neighbouring cells trough a sharing cell face. Pairs of indices in both permutations define nonzero matrix entries and matrix sparsity pattern. This format enables fast identifications of neighbouring cells for each cell from matrix sparsity pattern, namely, it is the list of indices of all non-zero matrix entries in a specific row, defined by cell number, except, of course, the diagonal cell, since it represents the specific cell in question. This often comes in handy when making local calculations pertinent only to cells' immediate surrounding.

In freeCappuccino, the csrMatrix object is defined from mesh data, to store sparsity pattern information and discretisation coefficients, using Fortran derived types:

```
type csrMatrix
  integer, dimension(:), allocatable :: ioffset
  integer, dimension(:), allocatable :: ja
  real(dp), dimension(:), allocatable :: coef
end type
```

By extending the csrMatrix type, we derive fvEquation derived data type. Finite volume equation is a datatype that besides sparse matrix, holds RHS vector, defined below as source vector and past values from two consecutive time-steps, o, and oo. This is done so the implicit Backward Differentiation of second order (BDF2 algorithm) can seamlessly shift time values during the simulation, and this information is pertinent to each field that is implicitly solved.

```
type, extends(csrMatrix) :: fvEquation
  real(dp), dimension(:), allocatable :: source
  real(dp), dimension(:), allocatable :: o
  real(dp), dimension(:), allocatable :: oo
end type fvEquation
```

Further extension by inheritance is done defining fvVectorEquation, where same data as above is hold for each vector component.

The fvEquation and fvVectorEquation have overloaded three basic arithmetic operations. The plus '+' sign is overloaded so one can add fvEquation and volScalarField, and do artihmetics with implicitly discretised tensor fields and volume field that define source terms in approximated governing equations. Overloading summation also applies when two objects of type e.g. fvEquation need to be added together. The same applies for minus sign, which is overloaded in the same way. What is interesting is the == sign, where basic idea is based on the one in Open-FOAM, i.e. this sign defines the same as the plus sign but is important for emulating the mathematical form of equations. The == sign is used to add volume source to linear equation systems' source if we have fvEquation and volScalarField objects respectively, on two sides of the sign, and it means adding two objects of type fvEquation if these are on different sides of the sign. That is principle of polymorphism in action. The solve function call accepts fvEquation object, solves linear system by an iterative algorithm and returns unknown volScalarField.

In freeCappuccino there are several built-in iterative solvers for sparse linear systems. The built-in solvers belong to two categories, stationary iterative methods such as Jacobi, and Gauss-Seidel, and preconditioned methods from Krylov family of algorithms - Diagonally Preconditioned Conjugate Gradient, Incomplete Cholesky Conjugate Gradient, ILU(0) preconditioned Bi-CGStab, and ILU(0) preconditioned GMRES [10]. All of these have also their parallel versions. Finally for wider choice of algorithms including Smooth Agglomeration-Algebraic Multigrid, freeCappuccino interfaces external library for sparse numerical linear algebra LIS [11].

6 Parallelisation

Modern problems in computational science and engineering often require resolution of many spatial and temporal scales which require highly advanced parallel application specific algorithms and high-performance computational (HPC) resources. State of the art HPC architectures available to researchers over the world offer combined power of multicore CPUs and GPUs with inherent parallelization approaches. Many core CPU architectures are arranged in multi node clusters offering greater computational power to the user. Fast growth in computational power available necessitates development of novel simulation approaches and tools which could use advantages of these resources in their full potential and still are able to maintain qualities such as portability and easy maintenance. State of the art codes should demonstrate capability of scaling with increasing the number of distributed computational nodes and computational cores. The problem they face and performance bottlenecks are result of the algorithmic and implementation choices. One typical example is that of numerical simulation of turbulent single/multiphase fluid flows. Common technical applications involve such flows in complex geometries where vortices at different scales are present. These vortices need high spatial numerical resolution in certain regions. Additionally to high spatial resolution, complex turbulent flows present multiscale features in time domain which lead to high demand in computational resources for long time integration over small discrete time-steps. Additional effects such as heat transfer, particulate transport and chemical reactions, or need for interphase interfaces capturing or resolving lead to stiff problems which further burden the simulation algorithm. Such problems require novel computational approaches which can follow the raise of physical complexity.

The approach adopted in freeCappuccino is based on domain decomposition and Single Program Multiple Data (SPMD) parallelisation paradigm implemented using Message Passing Interface (MPI). The required exchange of the data at certain points is achieved by a high level call to exchange routine in which data is moved to/from buffers and a call to MPI MPI_SENDRECV_REPLACE function as the main routine driving exchange of data is done. Other MPI function calls include MPI_ALLREDUCE global communication often employed in linear solvers to perform e.g. global residual norm calculation. At some places threading is used trough calls to OpenMP library.

7 Example Simulation Results

Here we illustrate use of freeCappuccino in most important context, as a fully capable computational fluid dynamics solver.

The first example is well known lid driven cavity case. The side of the cavity is d = 0.1, lid velocity is U_{lid} = 1 m/s, density is set to unit value and dynamic viscosity to μ = 0.01 Pas, which give Reynolds number Re = 10. The mesh is uniform and quite coarse, consisting of only 400 cells. The case setup details and computational mesh is provided as a basic tutorial example in OpenFOAM library, therefore this example also serves to test the ability of freeCappuccino to use meshes written in polyMesh format. Simulation is steady state, run in laminar mode, and SIMPLE algorithm is used to solve

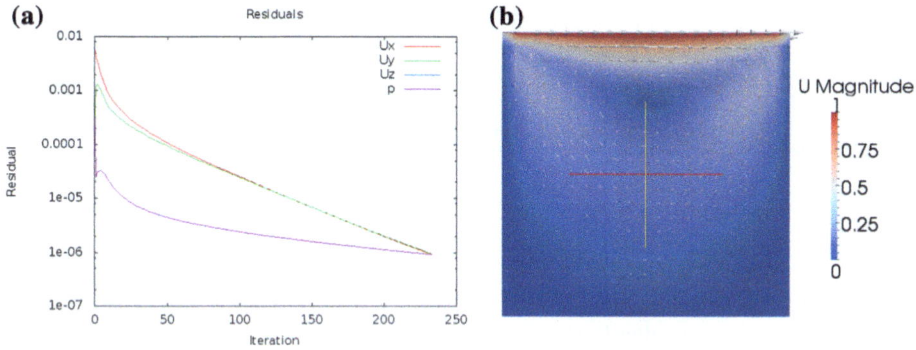

Fig. 2. Lid-driven cavity case. (a) Convergence of the SIMPLE algorithm, and (b) Velocity magnitude and flow velocity vectors at steady state.

coupling of velocity and pressure fields. Figure 2 shows simulation convergence and simulation result for velocity magnitude.

Next example is transient turbulent flow over backward facing step, also taken from OpenFOAM example set, known there as pitzDaily case. It can be used to validate several solver capabilities and as such is used as an example case for a few different OpenFOAM solvers. Here we are interested to show results of using SIMPLE in non-stationary mode. The mesh consists of 12225 cells. Turbulence is modelled using standard k-ε turbulence model. Convection scheme is second order upwind and Venkatakrishnan gradient limiter is employed. Time-step is set to $\Delta t = 0.001$ s, and number of SIMPLE iterations per time-step is limited to 30.

The flow develops to steady state after showing initial transients, which are shown in Fig. 3 and Fig. 4 for time $t = 0.02$ s. Other options also exist for non-stationary run besides running SIMPLE in transient mode, but this ability is useful because of the stability when larger time-steps are used. Time-stepping is based on backward differentiation formula of second order.

The code is extensively tested for turbulent flows in the atmospheric boundary layer of complex terrains. Most of the algorithmic developments were done to enable efficient treatment of highly distorted nonorthogonal meshes, which are often present

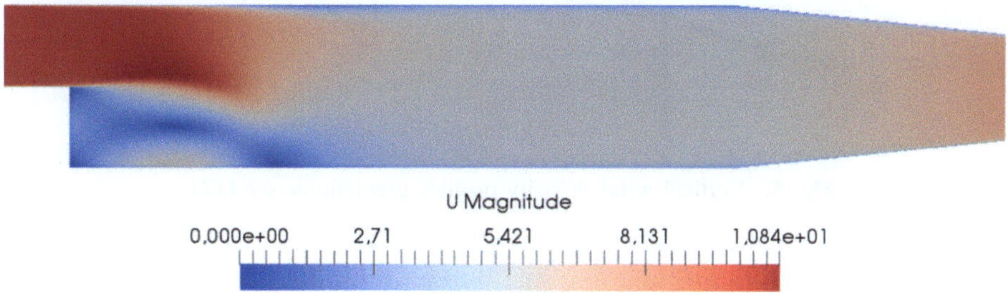

Fig. 3. Snapshot of initial transient flow development in backward facing step case of Pitz and Daily - Velocity magnitude

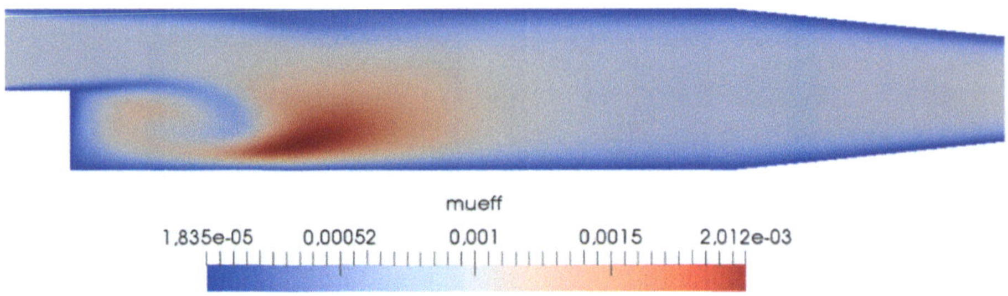

Fig. 4. Snapshot of initial transient flow development in backward facing step case of Pitz and Daily - Effective viscosity, after 0.02 s.

when real terrains are modelled. Major results are reported elsewhere e.g. [9, 12], and readers are encouraged to consult these references for more detail.

Here we show a comparison of vertical wind-velocity profiles over complex terrain, in wind-farm sitting study, Fig. 5. The differences between results of two codes with the same model are greater than between two different models in the same code. We believe that numerous algorithmic improvements aimed at flow over complex terrain cases had led to improved predictions in this case.

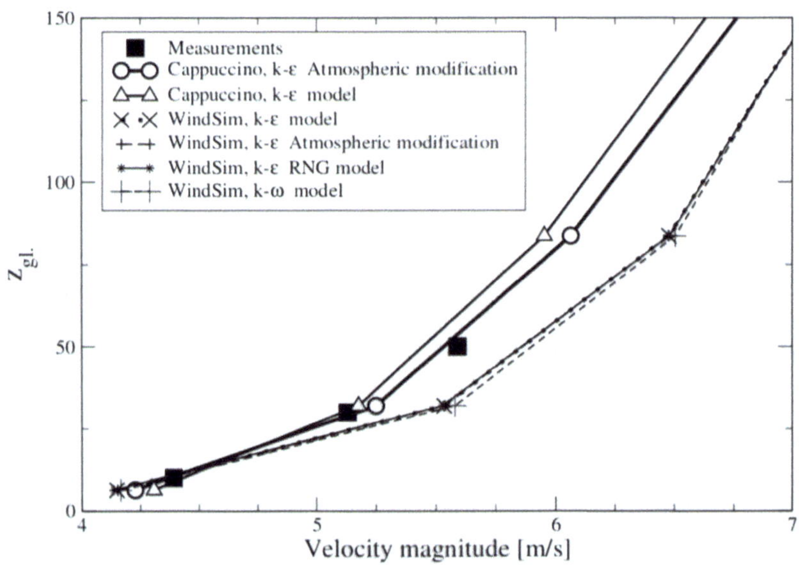

Fig. 5. Vertical wind velocity profiles predictions, see [12].

8 Conclusion

In this paper we have described a design approach for a computational science and engineering software library aimed at computational fluid dynamics and in general to computational continuum mechanics. Design decisions were such that any new solver should easily be implemented with provided tools for general unstructured mesh manipulation, operation over tensor fields and approximation of differential operators needed to write governing equations of the problem in continuum mechanics. Code parallelisation enables higher numerical efficiency for demanding problems. The code is provided in public repository with Git version control (www.github.com/nikola-m/freeCappuccino) with the hope to promote scientific collaboration and to serve as a basis for reproducibility of scientific results.

Acknowledgements. Support of the Ministry of Education, Science and Technological Development of republic of Serbia, trough project TR-33036 is greatly acknowledged.

References

1. Ince, D.C., Hatton, L., Graham-Cumming, J.: The case for open computer programs. Nature **482**, 485–488 (2012)
2. Ferziger, J.H., Perić, M.: Computational Methods for Fluid Dynamics, 2nd edn. Springer, Berlin (1999)
3. Weller, H.G., Tabor, G., Jasak, H., Fureby, C.: A tensorial approach to computational continuum mechanics using object oriented techniques. Comput. Phys. **12**(6), 620–631 (1998)
4. Jasak, H., Jemcov, A., Tuković, Ž.: OpenFOAM: a C++ library for complex physics simulations. In: International Workshop on Coupled Methods in Numerical Dynamics IUC, Dubrovnik, Croatia, 19th–21th September 2007
5. Vidanović, N., Rašuo, B., Kastratović, G., Maksimović, S., Ćurčić, D., Samardžić, M.: Aerodynamic-structural missile fin optimization. Aerosp. Sci. Technol. **65**, 26–45 (2017)
6. Bangerth, W., Heister, T.: What makes computational open source software libraries successful? Comput. Sci. Discov. **6**(1), 015010 (2013). http://stacks.iop.org/1749-4699/6/i=1/a=015010
7. Wang, Z.J.: Improved formulation for geometric properties of arbitrary polyhedra. AIAA J. **37**(10), 1326–1327 (1999)
8. Bartlett, R.E.: A simple convention for the specification of linear algebra function prototypes in C++. ACM Trans. Math. Softw. 1–6 (2015). https://trilinos.org/docs/dev/packages/thyra/doc/html/LinearAlgebraFunctionConvention.pdf
9. Mirkov, N., Rašuo, B., Kenjereš, S.: On the improved finite volume procedure for simulation of turbulent flows over real complex terrains. J. Comput. Phys. **287**, 18–45 (2015)
10. Saad, Y.: Iterative Methods for Sparse Linear Systems, 2nd edn. Society for Industrial and Applied Mathematics, Philadelphia (2003)
11. Library of Iterative Solvers for linear systems (LIS). http://www.ssisc.org/lisl
12. Mirkov, N., Stevanović, Ž.: New non-orthogonality treatment for atmospheric boundary layer flow simulation above highly non-uniform terrains. Therm. Sci. **20**(Suppl. 1), 223–233 (2016)

Multiparameter Structural Optimization of Pressure Vessel with Two Nozzles

Martina Balac(✉) and Aleksandar Grbovic

Faculty of Mechanical Engineering, University of Belgrade, Kraljice Marije 16, Belgrade, Serbia
mbalac@mas.bg.ac.rs

Abstract. Structural analysis of pressure equipment (vessels) has always been a huge challenge for researchers. Pressure vessels are usually subjected to different loads in exploitation and small defects can lead to failure of the equipment, which may result in loss of life, health hazards and damage of property. Modern approach of stress and strain analysis of the influence of welded nozzles on pressure vessels involves numerical and experimental testing. In this research, 3D Digital Image Correlation (DIC) method for analyzing full field surface strain and stress, including camera system in combination with Aramis software, was used. After determination of critical areas with highest von Mises stresses and strain concentrations, numerical analysis of equivalent 3D model was performed in Ansys Workbench software. The aim of this paper is to present detailed parameter optimization of pressure vessel with two nozzles based on finite element analysis (FEA) of the structure. Several geometrical parameters were varied to obtain the optimum geometry of the pressure vessel, capable of withstanding the service load without plastic deformation. It is shown that carried out optimization gives the minimum weight of pressure vessel with optimized wall and nozzle thicknesses for the given load.

Keywords: Pressure vessel · Digital Image Correlation method
Finite element analysis · Optimization · Response surface method

1 Introduction

Pressure equipment is the most widely used equipment in various industrial sectors and is usually subjected to different loads in exploitation. Many researches deal with stress and strain analysis of pressure equipment using both experimental and numerical methods [1–7]. Due to different technical and technological requirements, nozzles are often welded on vessel's shell producing geometrical discontinuity, which results in stress and strain increase in that area [2, 6, 7].

Regarding the appropriate structural design of the pressure vessel, the optimization approach attracted many researchers, both in practice and research [8–11]. Optimization of the location and size of opening (hole) in a pressure vessel cylinder was presented in the research of Hyder and Atif [12]. They analysed three thick-walled cylinders with internal diameter (20, 25 and 30 cm), having 30 cm height and wall thickness of 20 mm. Firstly, they conducted the analysis of pressure vessel cylinder

N. Mitrovic et al. (Eds.): CNNTech 2018, LNNS 54, pp. 148–158, 2019.
https://doi.org/10.1007/978-3-319-99620-2_12

without hole, calculating tangential, longitudinal, radial, and Von Misses stresses on cylinder. Secondly, the optimization of the hole size was carried out by drilling holes with diameters 4, 8, 10, 12, 14, 16 and 20 mm at the centres of all three thick cylinders. A method that directly minimizes the weight of the vessel by optimally determining the wall thickness is presented by Widiharso et al. [13].

Zhang and Yang showed in their paper [14] that modern optimization design combines mathematical programming with computer technology. They selected two geometrical parameters as design variables and the maximum volume of the pressure vessel as objective function. They found that different tolerances of design variables and constraints have effect on final design results. However, the design variables and the objective function are close to the optimal value along with the increase of iteration.

Bochare [15] showed that an appropriate location and size of the opening in a pressure vessel results in minimizing the stresses induced due to the stress concentration resulting from the end flanges and other attachments. The main aim of his work was to design and optimize the spherical and elliptical head profile with hole on the head. Analysis was done for elliptical and spherical heads without and with holes and by changing the diameters and distances between the holes.

Structural analysis of pressure vessel with and without reinforcement of nozzle for offset of 0, 8, 16, 24 and 32 inches from vertical centre line at central cross section with different inclination angles 0°, 15°, 30° and 45° was presented in [16]. Analytical and finite element analysis (FEA) of cylindrical pressure vessel with different vessel's head (hemispherical, standard ellipsoidal, dished shape and torispherical) has been presented in [17, 18]. Stress distribution in the pressure vessel was compared for diverse types of pressure vessel heads and it was concluded that stresses and deformation in hemispherical heads are the lowest.

Considering previous research papers, the aim of this research was to optimize several parameters that have high impact on stress and strain state of the pressure vessel with two nozzles.

2 Experimental Analysis

2.1 Basic Model

Experiments were performed on horizontal pressure vessel with the following dimensions: diameter of the vessel $D = 378.4$ mm, thickness of the vessels shell $e_a = 1.5$ mm, length of the vessel $L = 770$ mm. Pressure vessel was made of X5CrNi1810 material (EN 10088). On the cylindrical shell of the vessel, two nozzles – nozzle 1 DN 50 (Ø60.3 × 2.9 mm) and nozzle 2 DN32 DN 32 (Ø42.4 × 2.6 mm) were placed according to the appropriate EN standard. Nozzles were welded on the vessel and placed at an angle of 90° with respect to the longitudinal axis of the vessel. The nozzle's height influence on shell stress/strain state is negligible. Nozzles are placed at the minimum allowed distance, 98.3 mm, calculated according to the standard EN 13345-3 [6]. Experiment using Digital Image Correlation (DIC) method was conducted. This method [19–21] overcomes the limitations of conventional experimental methods and enables full-field displacement and strain measurement. At the beginning of 80's Peters and Ranson [22] first employed DIC. They assumed that there is a correlation in the intensity patterns of surface images taken before and after deformation. In the last decade, DIC technique has attracted full attention. DIC does not depend on the type of material or on the shape of the tested object and it has a wide range of applications, such as monitoring the strain and displacement in simple specimen [23, 24], or in the more complex structures [4, 25]. In this research, ARAMIS system developed by GOM has been successfully used to measure full strain field on the pressure vessel with two nozzles.

Equipment for experimental analysis consisted of previously mentioned optical system for 3D strain analysis (special sets of stereo cameras and lenses), software package Aramis, astand that allows the security and stability of sensors, devices to control the supply and image capture, PC systems and additional LED lighting (Fig. 1).

Fig. 1. Experimental setup, 1 - Stereo cameras; 2 - Vessel; 3 - Nozzle 1, DN50; 4 - Nozzle 2, DN32; 5 - Metal support; 6 - Pressure gauge; 7 - Illumination; 8 - Water pump connected to the pressure vessel

Areas of interest were defined to include the field between two nozzles and weld joint at the vessel's head. Parameters for basic Aramis system setup were: measuring volume 100 × 75, measuring distance 800 mm, camera angle 26°, calibration object CP20 90 × 72. Before starting the experiment, vessel's surface was prepared according to the tables in the instruction manual [26]. Several internal pressure values (0.5 MPa, 1 MPa and 1.5 MPa) were used in experiments.

2.2 Experimental Findings

The highest strain values on the pressure vessels obtained in experiments with the pressure 1.5 MPa are shown in Fig. 2.

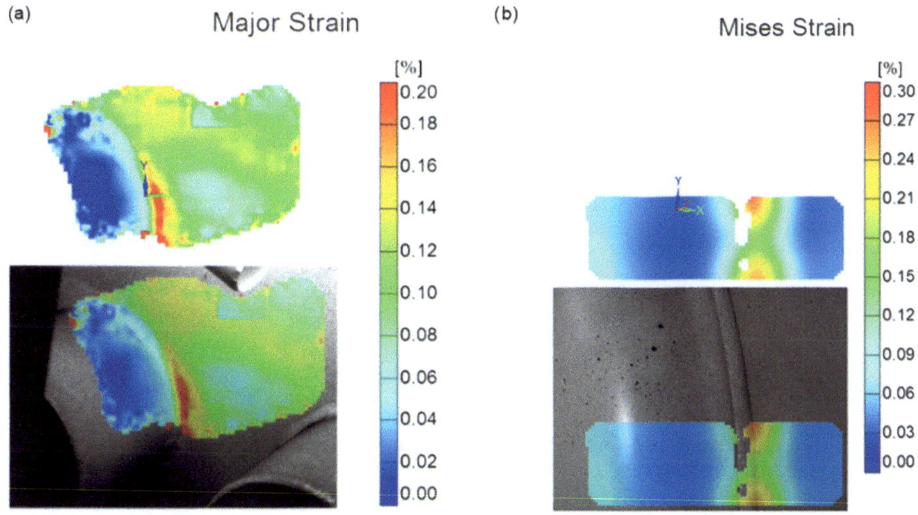

Fig. 2. Values of strain on pressure vessel for internal pressure of 1.5 MPa, (a) Strain distribution between the nozzles, (b) Strain on welded joint

Mises strain value near the nozzle 1 is 0.20% (Fig. 2a) and this area – where nozzle 1 is welded to the cylindrical surface – is the critical from the point of view of the crack initiation. Crack occurrence in this zone may lead to a complete failure of the vessel, as shown in [6, 27]. Figure 2b displays the strain distribution along the part of the weld joint of the vessel head. Value of the strain in that area varies from 0.10 to 0.25%, with dominant value of approximately 0.15%.

3 Finite Element Analysis

Based on the dimensions of the pressure vessel used in experiments, geometry of the vessel needed for numerical simulation was made in CATIA v5 software. Due to the symmetry of the vessel, only one quarter was designed and exported to Ansys Workbench software (Fig. 3). Applied boundary conditions and loads matched those used in experiments, while several different mesh densities had been used until the most appropriate was obtained. Final mesh consisted of 145,701 nodes and 72,732 solid elements. After applying pressure load, static nonlinear FEAs were carried out using material data obtained previously in the experiment with specimen [28]. The main purpose of these numerical simulations was verification of the developed model and – as it can be seen in Fig. 4 – strain values obtained in simulation were close to those obtained in the experiment. The highest strain values on outer surface appeared near nozzle 1 (Fig. 4a), in the same area where it appeared in experiment, with the values between 0.18% and 0.22% (max. value in experiment was 0.20%). Maximum strain in simulation 0.40% was observed on inner surface of a cylinder near nozzle 1 (Fig. 4b); however, this value couldn't be verified because no sensors have been used inside the pressure vessel. Nevertheless, values of strain measured on visible surfaces

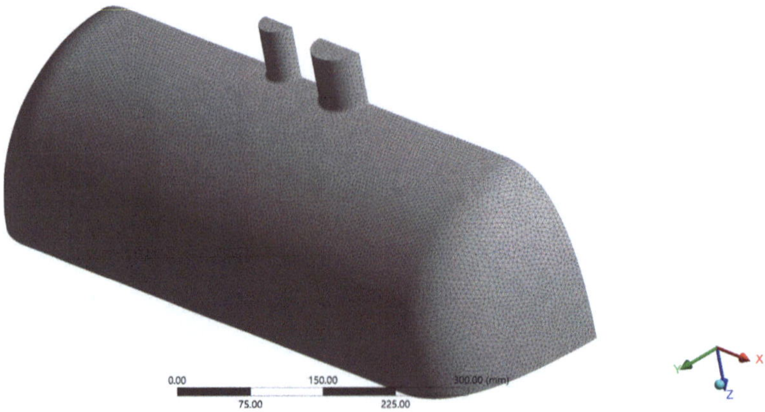

Fig. 3. Model of 1/4 of the vessel and finite element mesh used in FEA.

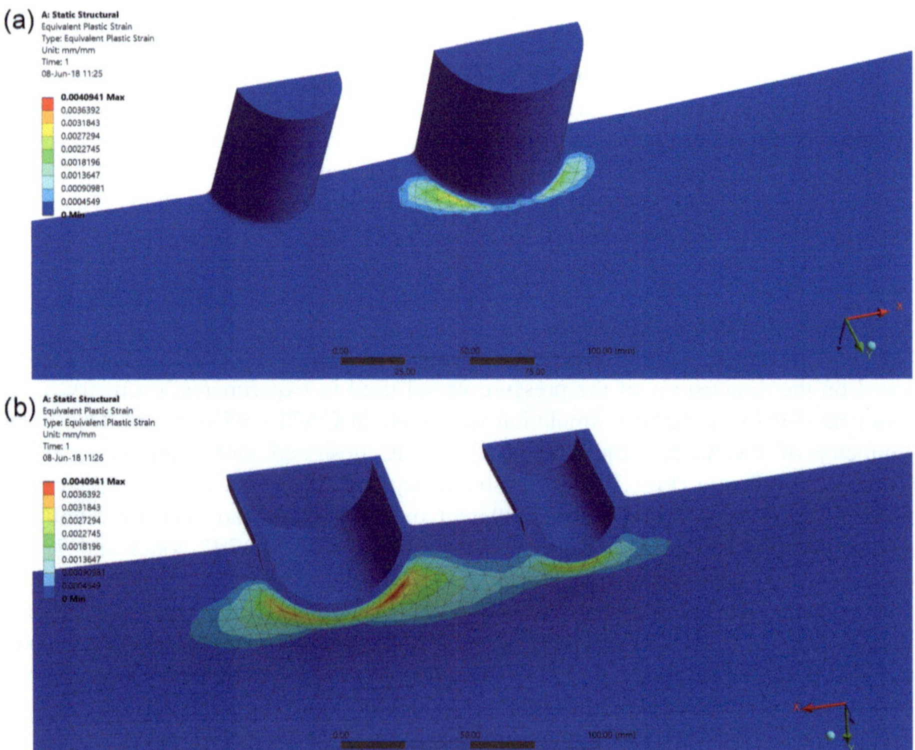

Fig. 4. Equivalent strain distribution on vessel for internal pressure of 1.5 MPa, (a) Strain near nozzle 1 is between 0.18 and 0.22%, (b) Max strain on inner surface

proved that numerical model was well defined. (It is worth mentioning that strain values near head's weld joint were about 0.15% and that maximum equivalent stress on vessel was around 200 MPa).

3.1 Optimization of the Pressure Vessel Geometry

After numerical model verification, next step in a research was optimization of the pressure vessel geometry with an aim of reducing its weight. Optimization design is developed in recent years and has been adopted by many researchers because it helps them to find the best solution from numerous design schemes. It combines mathematical programming with computer technology. In optimization design several steps should be followed: (1) Formation of parametric FEA file that includes unit types, inputting real constants, material parameters, verified solid model, mesh, constraints, loading conditions and required results; (2) Building of corresponding parameters' variables; (3) Executive optimization calculation, and (4) Testing and selecting the optimized results. Researches should be aware that there will be a series of feasible and infeasible design schemes throughout the process of optimization and that they must choose the best scheme and test the rationality of its results.

In a case of pressure vessel with two nozzles, four parameters – supposed to be the most influential on strain distribution under service load – were chosen: wall thickness of nozzle 1, wall thickness of nozzle 2, thickness of the vessel head and thickness of the cylindrical surface (Fig. 5). Upper and lower boundaries for each parameter have been selected after parameter choice. At the end of optimization process, the best parameter value should have been within defined boundaries.

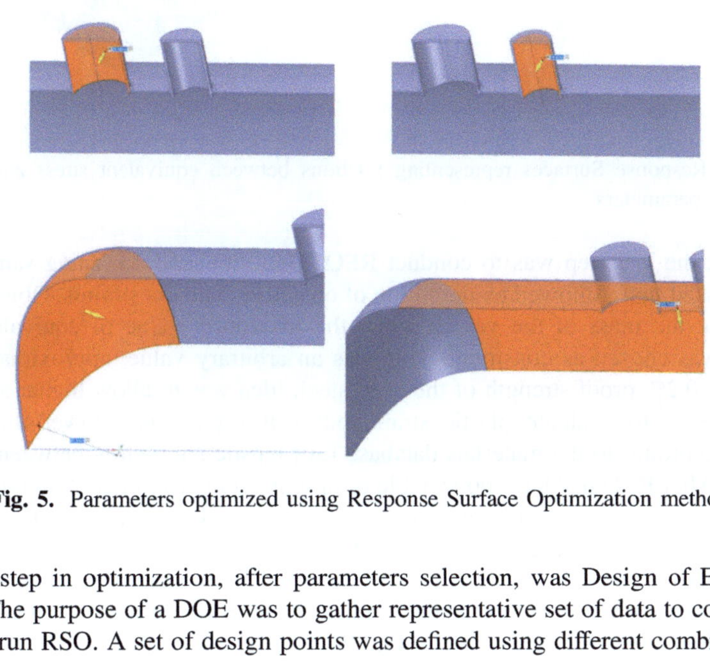

Fig. 5. Parameters optimized using Response Surface Optimization method

Next step in optimization, after parameters selection, was Design of Experiment (DOE). The purpose of a DOE was to gather representative set of data to compute RS and then run RSO. A set of design points was defined using different combinations of values for all four parameters within predefined boundaries. For that purpose, Central Composite Design (CCD) scheme was used. CCD is five-level fractional factorial design suitable for calibrating the quadratic response model [28]. There are 5 types of CCDs available in Ansys Workbench each with their own benefits and drawbacks: Auto Defined, Face Centred, Rotatable, VIF (Variance Inflation Factor) Optimality and G-Optimality. In optimization of pressure vessel Face Centred CCD was used.

It is important to emphasize that the RS accuracy greatly depends on the DOE scheme, and especially the number of design points that were calculated. In a DOE study, the amount of design points increases quickly as the number of input parameters increases, which can reduce efficiency of the analysis process. It is recommended to exclude unimportant input parameters from the DOE sampling in order to reduce time to solution. The parameter correlation tool allows researcher to identify unimportant parameters [29]. However, in pressure vessel optimization all four parameters were identified as important and 25 design points were generated. Calculation took about 120 min (I7 CPU at 3.5 GHz, 64 GB of RAM) and, as a result of good DOE, several response surfaces were obtained (Fig. 6).

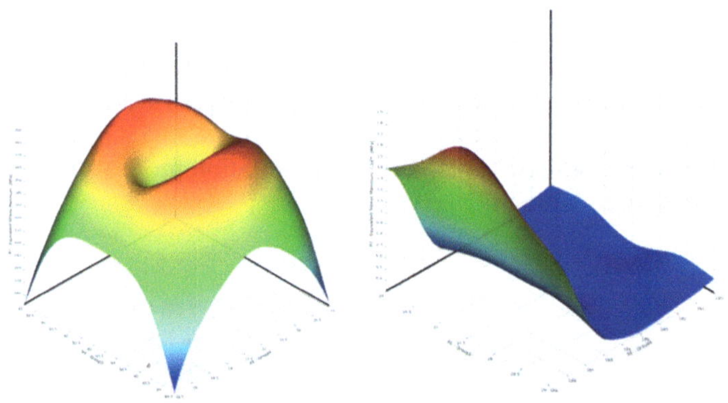

Fig. 6. 3D Response Surfaces representing relations between equivalent stress and different optimization parameters

Finally, the last step was to conduct RSO. One thousand screening samples had been generated first, followed by definition of objectives and constraints. Objective was to minimize the mass of the vessel, while the maximum value of equivalent stress 300 MPa was chosen as constraint. This was an arbitrary value, approximately 50% higher than 0.2% proof strength of the used steel. Idea was to allow higher stresses in pressure vessel to evaluate plastic strain, but still significantly lower than tensile strength; according to the materials database [30] tensile strength is between 500 and 750 MPa. After RSO had been done, table of optimization gave three "candidates" (i.e. combinations of parameters) based on specified goal and constraint. Trade off study was then conducted to reveal feasible points and, eventually, the best candidate was selected. Original model was updated with new values of parameters and numerical simulation was carried out once again. Values of strain in the area of interest on the optimized pressure vessel are given in Fig. 7.

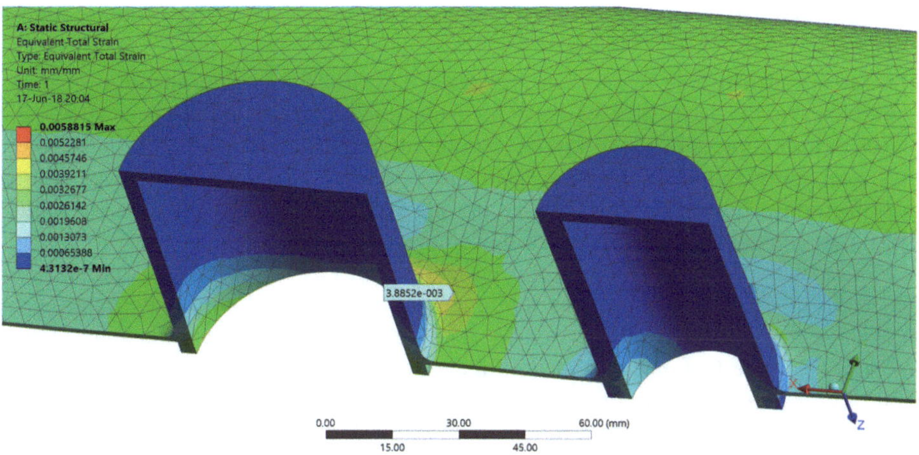

Fig. 7. Equivalent strain distribution on pressure vessel with optimized geometry

4 Discussion

Results obtained in experiment and FEA showed good agreement, i.e. both approaches indicated the same area of the highest strain in the vicinity of the one of the nozzles. This was the sign that all elements of the numerical model (geometry, load, boundary conditions) were well defined.

Response Surface Optimization (RSO) method was used to optimize geometry of the pressure vessel parts (shell, head and nozzles). Response surfaces (RS) are functions of different nature where the output parameters are described in terms of the input parameters. They provide the approximated values of output parameters everywhere in the analysed design space, without need to perform a complete solution. There are six RS types in Ansys Workbench, but in this research genetic aggregation (GA) type was used. GA runs an iterative genetic algorithm to find the best RS type and settings for each output parameter. It selects the best ones and combines them to build an "aggregation" of several RS. This results to the best RS quality and different settings for each output parameter. The goal is to meet 3 main criteria to obtain the best RS: accuracy, reliability and smoothness.

Even if the GA runs on several CPUs, this method can still be long when dealing with high number of design points and/or parameters (mesh density also influences the time to result). RSO methods are suitable for problems using up to 15 input parameters, but – due to the possibility of very long calculations – in this research 4 parameters were used as input and 2 as the output parameters. The objectives of research were to reduce the mass of the pressure vessel for given load and to keep the vessel below the predefined stress value.

5 Conclusion

The safety performance of pressure vessels is arousing increasing attention of many researches. In the past, they've been focusing on the stress and strain analysis and plastic deformation of pressure vessels. At present, the main aim is to solve the problem of shape optimization of the vessels in order to save materials and energy during manufacturing, but preserve reliability during exploitation. This can directly reduce costs and, consequently, the price of this equipment.

This paper presents a numerically based attempt in pressure vessel optimization. Objective of the optimization was minimization of the vessel's mass by varying the thicknesses of nozzles' walls as well as of vessel head and cylindrical surface. Using Response Surface Optimization method mass of the vessel was reduced from 3.642 to 3.383 kg, which gives the total reduction of 1.036 kg (considering that one fourth of the vessel was modelled). Thicknesses of nozzles' walls were increased by 0.85 mm (nozzle 1) and 3.2 mm (nozzle 2), while the thickness of the cylindrical surface was reduced by 0.3 mm (from 1.5 to 1.2 mm). Thickness of the vessel head remained almost the same: initial was 2.2 mm while final somewhat higher − 2.23 mm.

Figure 7 shows that strain value around nozzle 1 on optimized vessel is almost doubled (0.388% compared to 0.20% on original model and measured in experiment) which is expected considering the fact that thickness of cylindrical surface is now reduced, and the higher value of stress was allowed in optimization (new value of maximum equivalent stress is 315 MPa compared to 200 MPa on original vessel). Nevertheless, the main goal is achieved, and further investigations will show are these values of stress and strain acceptable (from the point of view of vessel integrity) and if they are, is there more room for further mass reduction. At least, additional methods of optimization can be tested and other parameters (like distance between the nozzles, for example) might be chosen.

Acknowledgements. Study presented in this paper is part of the Project TR 35031 financed by Ministry of Education, Science and Technological Development of Republic of Serbia.

References

1. Sang, Z.F., Xue, L.P., Lin, Y.J., Widera, G.E.O.: Limit and burst pressures for a cylindrical shell intersection with intermediate diameter ratio. Int. J. Press. Vessels Pip. **79**, 341–349 (2002). https://doi.org/10.1016/S0308-0161(02)00028-5
2. Petrovic, A., Balac, M., Jovovic, A., Dedic, A.: Oblique nozzle loaded by the torque moment: stress state in cylindrical shells on the pressure vessel. Proc. Inst. Mech. Eng. Part C J. Mech. Eng. Sci. **226**(3), 567–575 (2012). https://doi.org/10.1177/0954406211415907
3. Bajic, D., Momcilovic, N., Maneski, T., Balac, M., Kozak, D., Culafic, S.: Numerical and experimental determination of stress concentration factor for a pipe branch model. Tehnickivjesnik Tech. Gaz. **24**(3), 687–692 (2017). https://doi.org/10.17559/tv-20151126222916
4. Mitrovic, N., Petrovic, A., Milosevic, M., Momcilovic, N., Miskovic, Z., Maneski, T., Popovic, P.: Experimental and numerical study of globe valve housing. Chem. Ind. **71**(3), 251–257 (2017). https://doi.org/10.2298/hemind160516035m

5. Maneski, T., Bajic, B., Momcilovic, N., Milosevic, M.V., Balac, M.: Determination of internal pressure value causing pipe branch model to plastically deform. FME Trans. **46**(2), 218–223 (2018). https://doi.org/10.5937/fmet1802218m
6. Balac, M., Grbovic, A., Petrovic, A., Popovic, V.: Fem analysis of pressure vessel with an investigation of crack growth on cylindrical surface. Eksploatacjai Niezawodnosc Maint. Reliab. **20**(3), 378–386 (2018). https://doi.org/10.17531/ein.2018.3.5
7. Lu, M.H., Yu, J.S., Chen, J.J.: The effect of analysis model on the stress intensity calculation for the nozzle attached to pressure vessel under internal pressure loading. Int. J. Press. Vessels Pip. **117–118**, 9–16 (2014). https://doi.org/10.1016/j.ijpvp.2013.10.004
8. Sathe, A.A., Maurya, V.R., Tamhane, S.V., Save, A.P., Nikam, P.V.: Design and analysis of pressure vessel components as per ASME Sec. VIII Div. III. Int. J. Eng. Dev. Res. **6**(1), 834–840 (2018)
9. Sahu, Y.K., Nagpal, S.: Design and optimization of a low pressure vessel. Res. J. Eng. **6**(7), 1–6 (2017)
10. Carbonari, R.C., Muñoz-Rojas, P.A., Andrade, E.Q., Paulino, G.H., Nishimoto, K., Silva, E. C.N.: Design of pressure vessels using shape optimization: an integrated approach. Int. J. Press. Vessels Pip. **88**, 198–212 (2011). https://doi.org/10.1016/j.ijpvp.2011.05.005
11. Kulkarni, A.A., Jatkar, K.H.: A review on optimization of finite element modelling for structural analysis of pressure vessel. Int. J. Eng. Trends Technol. (IJETT) **12**(1), 20–22 (2014)
12. Javed Hyder, M., Asif, M.: Optimization of location and size of opening in a pressure vessel cylinder using ANSYS. Eng. Fail. Anal. **15**, 1–19 (2008). https://doi.org/10.1016/j.engfailanal.2007.01.002. ISSN 2231-5381
13. Widiharso, H.S., Tauviqirrahman, M., Jamari, M.J.: Thickness optimization of pressure vessel for minimum weight using finite element method (FEM). Int. J. Eng. Technol. **8**(6), 2676–2682 (2017). https://doi.org/10.21817/ijet/2016/v8i6/160806225
14. Zhang, C., Yang, F.: Pressure vessel optimization design based on the finite element analysis. Appl. Mech. Mater. **65**, 281–284 (2011). https://doi.org/10.4028/www.scientific.net/AMM.65.281
15. Bochare, S.: Optimization of location and size of opening in a pressure vessel cylinder for spherical and elliptical head. Int. J. Emerg. Technol. **6**(2), 229–234 (2015)
16. Gupta, S.R., Desai, A., Vora, C.P.: Optimize nozzle location for minimization of stress in pressure vessel. Int. J. Adv. Eng. Res. Dev. (IJAERD) **1**(6), 1–14 (2014)
17. Digvijay, K., Jewargi, S.S.: Stress analysis of pressure vessel with different type of end connections by FEA. Int. J. Innov. Res. Sci. Eng. Technol. **4**(5), 2769–2775 (2015). https://doi.org/10.15680/IJIRSET.2015.0405016
18. Lawate, S., Deshmukh, B.B.: Analysis of heads of pressure vessel. Int. J. Innov. Res. Sci. Eng. Technol. **4**(2), 759–765 (2015). https://doi.org/10.15680/ijirset.2015.0402098
19. Orteu, J.J.: 3-D computer vision in experimental mechanics. Opt. Lasers Eng. **47**, 282–291 (2009). https://doi.org/10.1016/j.optlaseng.2007.11.009
20. Sedmak, A., Milosevic, M., Mitrovic, N., Petrovic, A., Maneski, T.: Digital image correlation in experimental mechanical analysis. Struct. Integr. Life **12**, 39–42 (2012)
21. Kawabayashi, H., Matsuda, H., Zhao, C., Konzuma, H., Yamashita, T.: Modeling cylindrical shell with initial imperfections by using optical method. In: The Fifth China: Japan Joint Seminar for the Graduate Students in Civil Engineering, Shanghai, China, pp. 181–186 (2008)
22. Peters, W.H., Ranson, W.F.: Digital imaging techniques in experimental stress analysis. Opt. Eng. **21**(3), 427–431 (1982). https://doi.org/10.1117/12.7972925

23. Rodrigues, L.D., Freire, J.L., Vieira, R.D., Castro, J.T.P.: Strain analysis of thin pipe pressure vessels using digital image correlation. J. Mech. Eng. Autom. **4**(2), 63–72 (2014). https://doi.org/10.5923/j.jmea.20140402.03
24. Chen, F., Chen, X., Xie, X., Feng, X., Yang, L.: Full-field 3D measurement using multi-camera digital image correlation system. Opt. Lasers Eng. **51**, 1044–1052 (2013). https://doi.org/10.1016/j.optlaseng.2013.03.001
25. Revilock, D.M., Thesken, J.C., Schmidt, T.E., Forsythe, B.S.: 3D digital image correlation of a composite overwrapped pressure vessel during hydrostatic pressure tests. In: Proceedings of the 2007 SEM Annual Conference and Exposition on Experimental and Applied Mechanics, Springfield, MA, pp. 1–15 (2007)
26. https://support.gom.com. Accessed 21 Nov 2016
27. Balać, M., Grbović, A., Petrović, A.: A numerical predictions of crack growth in a pressure vessel with welded nozzles. Struct. Integr. Life **15**(1), 55–61 (2015)
28. Balac, M.: Nozzles interaction influence on stress state and strains on cylindrical shell of the pressure vessel. Ph.D. thesis, University of Belgrade, Faculty of Mechanical Engineering (2014)
29. www.ansys.com. Accessed 15 Apr 2018
30. http://www.steelnumber.com/en/steel_composition_eu.php?name_id=100. Accessed 21 Mar 2018

Mathematical Modelling and Performance Analysis of a Small-Scale Combined Heat and Power System Based on Biomass Waste Downdraft Gasification

Marta Trninic[1(✉)], Dusan Todorovic[1], Aleksandar Jovovic[1], Dragoslava Stojiljkovic[1], Øyvind Skreiberg[2], Liang Wang[2], and Nebojsa Manic[1]

[1] Faculty of Mechanical Engineering, University of Belgrade, Kraljice Marije 16, 11020 Belgrade, Serbia
mtrninic@mas.bg.ac.rs
[2] SINTEF Energy Research, Postboks 4761 Sluppen, 7465 Trondheim, Norway

Abstract. The paper presents a simple mathematical model for designing, optimizing and simulating small–medium CHP scale plant with use of biomass waste downdraft gasification. A downdraft gasifier has been used as the starting point in the study, due to its low tar content and effective way of using heat in the engine's exhaust gases to dry and pyrolyze the different solid biomass waste. Hot water from the cooling circuit of the engine and from producer gas cooling is directly used for the district heating network, air or steam preheating. The mathematical model includes modelled components as a downdraft gasifier, an internal combustion engine using the characteristic equation approach method. The mathematical model enables the outputs of the plant to be evaluated and calculated for different types of biomass and operating conditions. The results demonstrate that it is a useful tool for assessing the performance of CHP plants using several types of biomass waste and enables comparisons to be made between operating conditions for real applications.

Keywords: Biomass · Downdraft gasification · CHP

1 Introduction

Gasification process is one of the most advanced and highly efficient processing routes to convert biomass into a useful gases and chemicals [1, 2]. The conversion is achieved, at high temperature (850–1500 °C), by reactions between a feed gas (air, enriched air, steam, oxygen) and biomass. The resultant mixture of gases (producer gas) formed during gasification process contains carbon monoxide (CO) and hydrogen (H_2), carbon dioxide (CO_2), methane (CH_4), water (H_2O), and a large panel of hydrocarbons including tars, and inorganic impurities in lower concentration [3–5]. Producer gas can be upgraded to synthesis gas (syngas) by adjusting the H_2/CO ratio and converted to any hydrocarbon [6]. It can be methanized or converted into other gaseous fuels (dimethyl-ether, hydrogen) or liquefied and upgraded to Fischer-Tropsch

N. Mitrovic et al. (Eds.): CNNTech 2018, LNNS 54, pp. 159–173, 2019.
https://doi.org/10.1007/978-3-319-99620-2_13

diesel, ethanol, or methanol [6]. Further, cleaned (free from tar, ash, alkali compounds) producer gas can be used in internal combustion engines, gas turbinesor fuel cells for power generation [7]. Integrating heating and electricity subsystems into a conventional plant could increase the plant's efficiency to 90% and to reduce its negative impact on the environment [4, 8].

One of efficient ways to integrate gasification with heating and electricity subsystems is by use of internal combustion (IC) gas engines. IC are best suited for small and medium scale plants from 1 to 10 MW [4, 9]. Also, IC gas engines have benefits like low capital cost, reliability, good part-load performance, high operating efficiency, and modularity and is quite safe to use [10].

The aim of this study is to develop simple and reliable simulation tools (mathematical model) to give a better understanding of the whole process of biomass gasification coupled with IC gas engine and to be used as preliminary tools to evaluate the characteristics of CHP biomass gasification plants. The mathematical model must be able to predict the performance of the whole system under varying operating conditions: different biomass characteristics, ambient temperature, gasifying agent, etc. This model should be useful, at a design stage, to evaluate the outputs of the plant for different types of biomass and operating conditions. In addition, the energetic performance of the CHP should be clearly stated.

This paper presents mathematical model development of a small-scale combined heat and power system, based on biomass waste downdraft gasification and IC gas engine, powered by corn cobs (as a form of waste biomass). With this model is possible to simulate how the heat from the producer gas and IC gas engine can be used to increase the performance of the system, for example by powering the gasification process (preheating air or generate steam) and heating water for district heating network (DH).

2 Development of Mathematical Models

Due to the inherent complexity of biomass gasification processes, modelling for simulation and prediction of performance of the processes is still an incipient activity [11]. In this paper, only description of gasification models is presented, as generally the simulation of equipment and processes after gasifier doesn't present a challenge. Different kinds of models have been proposed in order to explain the gasification process, with an interest towards the design, simulation, optimization, and process analysis of gasifiers. Regard to analysis of gasification process, the models can be divided into kinetic, thermodynamic equilibrium, artificial neural network models, Aspen Plus gasification models and computational fluid dynamics simulation models. In literature can be found a detailed review of recent biomass gasification models [2, 11–16]. In this paper, is presented the gasification model, part of developed downdraft CHP system model, which is no longer is considered as a black box. It involves main gasification sub-processes (drying, pyrolysis, gasification) and their products. The developed model, based on thermodynamic equilibrium calculations, has been validated with experimental published data of other authors. It, provides the opportunity to evaluate downdraft gasification processes as well as effects of variations in biomass properties and operating conditions. In further text, the development of whole gasification CHP system is presented.

3 Downdraft Gasification CHP System Description

Proposed configurations for the small-scale gasification plant contain the following components: a downdraft gasifier, an internal combustion (IC) gas engine (which is the prime mover of the system), heat exchangers for heat recovery and a gas clean-up section (Fig. 1).

The chemical energy stored in biomass, in the downdraft gasifier at 950 °C, is converted into the energy of a producer gas (mixture of N_2, H_2, CO, CO_2 and CH_4). Part of the biomass energy content is lost in the conversion process, both as heat loss and as energy stored in the charcoal [17]. After gasification process, producer gas exit downdraft gasifier at temperature around 500 °C [18]. Before entering the cleaning system, the producer gas (at 500 °C) needs to be cooled (up to 150 °C). The rejected heat can be used to pre-heat air and/or generate steam for the gasification, or to produce hot water for the DH [17–19]. The cooled producer gas passes through a gas cleaning system (e.g. cyclone for large solid particles removal, catalytic tar cracker for tar reduction, a bag filter for small particles and condensed tar removal) where is additionally cooled to 25 °C [18]. Afterwards, the cooled and cleaned producer gas is burned in IC gas engine to produce 320 kW of electrical power. Heat from exhaust gases and from the engine (oil and cooling water) is partially recovered and used to produce hot water for the DH.

4 CHP System Modelling

4.1 Process Model Simulator

The "Engineering Equation Solver (EES)" [20] has been found to be very suitable for modeling this kind of system, because it contains all of the necessary thermodynamic functions and it is possible for the model builder to make a user interface [2, 18].

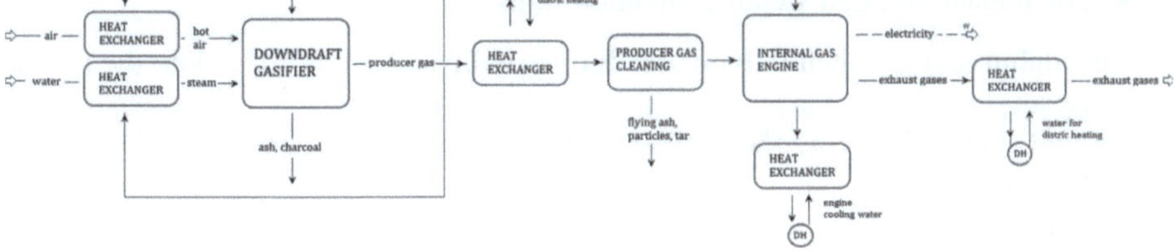

Fig. 1. The block scheme of the typical components of a small-scale gasification plant

4.2 Model Settings

Biomass Fuel

Corn cob, as a form of biomass waste is chosen as a feed into the downdraft gasifier. Proximate and ultimate analyses of corn cob were shown in works of Wang, Trninić et al. [21] and Trninić et al. [22]. The proximate and elemental analysis of corn cob composition is presented in Table 1.

Table 1. Proximate and elemental analysis of corn cob

Elemental analysis (wt%)[a]				
C	H	N	O[b]	S
47.61	6.27	0.55	43.89	0.23
Proximate analysis (wt%)[a]				
Moisture content[c]	VM	fix-C	ASH	HHV (MJ kg^{-1})
5.18	81.08	17.47	1.45	18.63

a - dry mass basis, b - by difference, c - as received

General Assumptions

The following assumptions were made:

1. Heat losses in pyrolysis and gasification units are estimated by the user as a percentage of biomass energy input to the system [2].
2. Corn cobs are assumed to enter the CHP plant at 25 °C and 1 atm.
3. The air for the gasification process is considered as dry, containing only: 21% O_2, 78% N_2 (volume fraction).
4. Modelling of the IC gas engine was carried out without consideration of the thermodynamic cycle and mechanical aspect analysis.
5. The gasification consists of a series of sub-processes:
    - Drying unit, that predicts the removal of moisture from raw biomass. The percentage of removed moisture can alternatively be set by the user.
    - Pyrolysis unit that, using empirical correlations, predicts the formation of pyrolysis products (charcoal and volatiles, including tar).
    - Gasification unit, that predicts the formation of gasification products (gas, including small amount of charcoal and tar).
    - Air preheating, and steam generation units.
6. All sulphur and nitrogen in biomass, during pyrolysis, is converted into the tar and charcoal.
7. Tar and charcoal leaving the gasifier as a percentage of tar and charcoal produced in the pyrolysis unit [2].
8. Particles leaving the gasifier are set by the user as mg/Nm3 in the producer gas. These particles are considered to consist only of carbon.
9. Gas products consists of CO_2, CO, H_2, CH_4, N_2, and H_2O.
10. Setting the amount of CH_4 produced.

11. All Sulphur in biomass is converted into the ash.
12. The model considers that producer gas completely cleaned from particles, tar and organic and inorganic impurities (through a water scrubber cyclone, bag filter etc.).

4.3 Model Description of the Different CHP Units

Gasifier Model

The gasification model consists of a series of sub-processes, each containing one process (biomass drying, pyrolysis, gasification, air preheating, and steam generation), see Fig. 2.

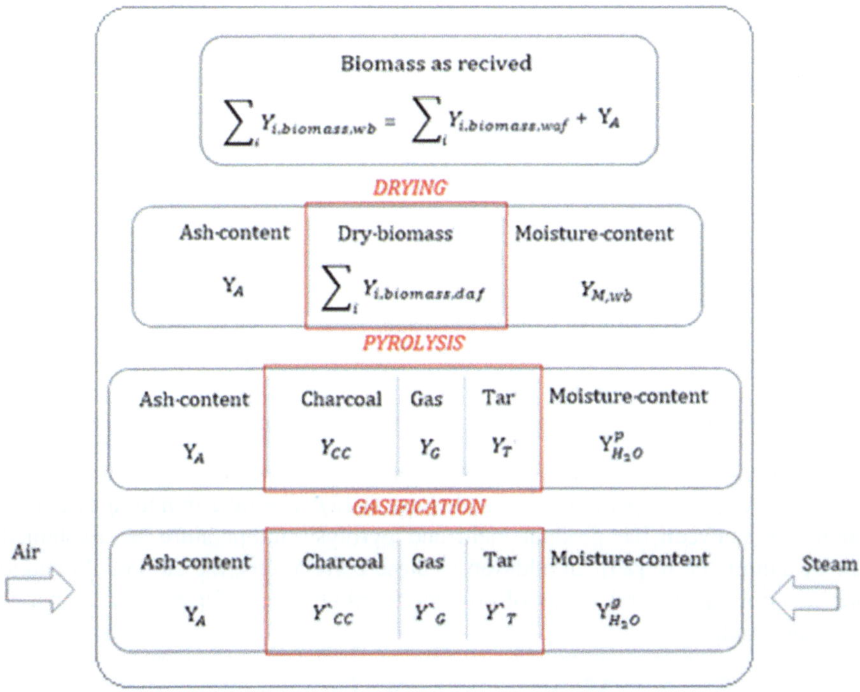

Fig. 2. Overall mass balance for the biomass gasification process

Gasification sub-processes can be described with the equations:

Drying:

$$\sum_i Y_{i,biomass,waf} + Y_A \underset{drying}{\rightarrow} \sum_i Y_{i,biomass,daf} + Y_{M,wb} + Y_A \quad (1)$$

Pyrolysis [23]:

$$\sum_i Y_{i,biomass,daf} + Y_{M,wb} \underset{pyrolysis}{\rightarrow} \sum_i Y^p_{j,products,daf} + Y^p_{H_2O} + Y_A \quad (2)$$

Gasification:

$$\sum_i Y^p_{j,products,daf} + Y^p_{H_2O} + Y_A + Y_{AIR} + Y_{STEAM} \underset{gasification}{\rightarrow} \sum_i Y^g_{j,products,daf} + Y_A + Y^g_{H_2O} \quad (3)$$

where:

$\sum_i Y_{i,biomass,waf}$ is the mass fraction of the ith element (carbon, hydrogen, oxygen, nitrogen) in wet biomass on an ash free basis.
$\sum_i Y_{i,biomass,daf}$ is the mass fraction of the ith element (carbon, hydrogen, oxygen, nitrogen) in dry biomass on an ash free basis.
Y_A is the ash content in biomass on dry basis.
$Y_{M,wb}$ is the moisture content of biomass on dry ash free basis.
$\sum_i Y^p_{j,products,daf}$ is the mass fraction of the jth pyrolysis product (charcoal, tar and gas) on a dry ash free basis.
$Y^p_{H_2O}$ is the moisture content in the gas obtained in the pyrolysis process.
Y_{AIR} is the mass fraction of air.
Y_{STEAM} is the mass fraction of steam.
$\sum_i Y^g_{j,products,daf}$ is the mass fraction of the jth gasification product (charcoal, tar and gas) on a dry ash free basis.
$Y^g_{H_2O}$ is the moisture content in the gas obtained in the gasification process.

For prediction of pyrolysis products, empirical relationships between the product yield and pyrolysis temperature are used (Eqs. 4–12). The determination of empirical relationships between the product yield and pyrolysis temperature are explained in detail by Trninić et al. [23]. In addition to these correlations, the empirical equations, which describe the general trends of product distribution as a function of temperature, are set and used.

$$\sum_i Y^p_{j,products,daf} = Y_{cc} + Y_T + Y_G \quad (4)$$

$$Y_G = Y_{CO2} + Y_{CO} + Y_{CH4} + Y_{H_2} \quad (5)$$

Temperature dependent charcoal, tar and gas yields are given by [23].

$$Ycc = 7.97T^2 \cdot 10^{-5} - 0.125 \cdot T + 68.87 \quad (6)$$

$$Y_T = -1.38T^2 \cdot 10^{-4} + 0.12 \cdot T + 12.64 \quad (7)$$

$$Y_G = 1.12T^2 \cdot 10^{-4} - 0.058 \cdot T + 30.77 \quad (8)$$

Dependence of gas yield on pyrolysis temperature is described by:

$$Y_{CO} = -2.65T^2 \cdot 10^{-4} + 0.27 \cdot T - 32.71 \quad (9)$$

$$Y_{CO2} = -2.85T^2 \cdot 10^{-5} - 0.029 \cdot T + 70.89 \quad (10)$$

$$Y_{CH4} = 6.69T^2 \cdot 10^{-5} - 0.037 \cdot T + 4.28 \quad (11)$$

$$Y_{H2} = 7T^2 \cdot 10^{-5} - 0.0371T + 5.11 \quad (12)$$

In addition to these correlations, the energy, mass, and molar balances for each element (C, H, O, and N) are set and used to calculate the gasification products. An initial gasification temperature is assumed in the iterative solution procedure.

Model (operating) parameters (drying temperature, percentage of removed moisture, pyrolysis temperature, air inlet temperature, steam inlet temperature, gasification temperature and percentage of charcoal, tar and particles leaving the gasifier) can be directly introduced by the user.

The model predicts the producer gas yield, composition (volume fraction in % of CO, CO_2, CH_4, H_2, N_2 and H_2O) and heating value for a particular biomass with a specific ultimate composition and moisture content.

The results provided by this model were validated by comparison of results given by variation of different parameters with the experimental results from Senelwa [24] and Da Silva [25]. The results form model validation showed that the model is accurate (RMSD = 0.026 [5]) enough to predict the behavior of downdraft gasifiers and it is proved to be sensitive enough to evaluate the influence of equivalence ratio, air preheating, steam injection, oxygen enrichment and biomass moisture content on the quality of producer gas.

Producer Gas Cooling

The producer gas cooling is modelled with a Heater block 1. The temperature is lowered from 500 °C to 150 °C. It is assumed that producer gas after passing the gas cleaning unit is additionally cooled to 25 °C. The exhaust gas cooling is modelled with a Heater block 2. The temperature is lowered from 180 °C to 85 °C.

Internal Combustion Engine Model

The exhaust gases from IC gas engines are complex mixtures consisting principally of the products of complete combustion, small amounts of the oxidation products of sulphur and nitrogen, and compounds derived from the fuel and lubricant [26].

The exhaust gas composition is calculated based on the combustion stoichiometry. In the analytical model presented for the adiabatic combustion in this section it is assumed that all gases are ideal gases and their enthalpies and specific heats only change with temperature.

The electricity and heat generated by the IC gas engine are calculated based on the electrical (η_{el}) and thermal (η_{th}) efficiencies of a GE's Jenbacher JMS 208 GS-B.L gas engine ($\eta_{el} = 35.8\%, \eta_{th} = 41.9\%$) [27]. Electrical and thermal efficiencies are defined as follows [4]:

$$\eta_{el} = \frac{W}{LHV_{gas} \times V_{gas}} \quad (13)$$

$$\eta_{th} = \frac{Q}{LHV_{gas} \times V_{gas}} \quad (14)$$

The amount of corn cob supplied to the CHP plant is adjusted to have an electrical output of 320 kW. Exhaust gas temperature at full load is 500 °C [27]. A heater block is placed on the exhaust gas stream to recover heat from 500 °C to 85 °C for district heating.

The Efficiency of CHP System Performance

The energetic performance of the CHP plant can be evaluated, according to Francois et al. [4], from cold gas efficiency and electrical, thermal and CHP efficiencies.

The cold gas efficiency (η_{cge}) is defined as the ratio of the energy contained in producer gas (Q_{pgas}) to the energy contained in biomass fuel ($Q_{biomass}$) [4]:

$$\eta_{cge} = \frac{Q_{pgas}}{Q_{biomass}} = \frac{V_{pgas} \times LHV_{pgas}}{m_{biomass} \times LHV_{biomass}} \quad (15)$$

The electrical (η_e), thermal (η_t) and overall efficiencies (η_{CHP}) of a CHP plant are calculated as follows [4]:

$$\eta_e = \frac{W_e}{Q_{biomass}} \quad (16)$$

$$\eta_t = \frac{Q_{DH} + Q_{steam} + Q_{AIR}}{Q_{biomass}} \quad (17)$$

$$\eta_{CHP} = \frac{W_e + Q_{DH} + Q_{steam} + Q_{AIR}}{Q_{biomass}} \quad (18)$$

where, W_e (kW) is the electrical output of the IC gas engine, Q_{DH} (kW) is the heat provided to the District heating, Q_{steam} (kW) the heat provided for steam generation and Q_{AIR} (kW) the heat provided for the air preheating.

5 Results and Discussion

In the present paper, a scenario analysis of the gasification CHP plant configuration is presented.

5.1 Configuration

The modelled characteristics of gasification plant coupled with IC gas engine is presented in Table 2. The input was considered to be 239 kg/h of dry corn cob with 5.8% of moisture. The air is not preheated. In the model, the pyrolysis and gasification unit temperature has been adjusted to 450 °C and 950 °C respectively. Percentage of charcoal and tar leaving the pyrolysis and gasification unit is given to the model. It was assumed that after gasification, 5% of pyrolysis charcoal and 5% of pyrolysis tar leaves the gasifier. Also, the CH4 percentage leaving the gasifier is also given to the model (3%). The temperature of the producer gas leaving the gasifier is considered to be 500 ° C. Table 3 shows the simulated results for produced gas. It can be seen that the model results are in good agreement with those reported by Senelwa [24] and Da Silva [25]. For the validation of the model, the temperature of the gasification process is settle to be 930 °C and 955 °C. Also, the percentage of CH4 is settle to be 2.5% and 3%. Heat is recovered from producer gas, exhaust gas and IC gas engine. Heat from producer gas is first used for possible air or steam preheating. Producer gas enters the Heater block 1, where temperature of the gas is lowered from 500 °C to 150 °C. The 83.75 kW of heat is used to heat water from 70 °C to 85 °C for District heating. Heat for District heating is also recovered from IC gas engine with 316.36 kW and exhaust gas cooling, Heater block 2, with 35.63 kW. The CHP plant provides 329.70 kW of electricity and 449.08 kW of heat for DH.

In the configuration studied, the cold gas efficiency (η_{cge}) is 85.35% when considering the energy content of clean producer gas. With reference to the corn cob energy content, the electrical (η_e), thermal (η_t) and overall (η_{CHP}) efficiencies of the CHP plant are 28.73%, 39.13% and 67.86% respectively.

5.2 Configuration

The modelled characteristics of gasification plant coupled with IC gas engine is presented in Table 2. The input was considered to be 222 kg/h of dry corn cob with air is preheated up to 400 °C. Other input is set up like in pervious case (temperature of pyrolysis and gasification process, percentages of charcoal and tar etc.). The temperature of the producer gas leaving the gasifier is considered to be 500 °C. Table 2 shows the simulated results for produced gas.

The CHP plant provides 330.40 kW of electricity and 423.56 kW of heat for District heating. Heat from producer gas, exhaust gas and IC engine is used for air preheating and water heating for DC. After leaving air preheating unit, temperature of the producer gas is lowered from 500 °C to 325 °C. Cooled producer gas enters the Heater block 1, where temperature of the gas is lowered from to 150 °C. The 39.62 kW of heat is used to heat water from 70 to 85 °C for District heating. Heat for District heating is also recovered from IC gas engine with 317.03 kW and exhaust gas cooling, Heater block 2, with 32.44 kW. In the configuration studied, the cold gas efficiency (η_{cge}) is 86.89% when considering the energy content of clean producer gas. With reference to the corn cob energy content, the electrical (η_e), thermal (η_t) and overall (η_{CHP}) efficiencies of the CHP plant are 29.69%, 39.07% and 68.76% respectively.

Table 2. Technical specifications of the gasification plant

CPH power plant-downdraft gasification with IC gas engine specification			
Fuel characteristics			
Fuel	Corn cob		
Size	Do = 10–20 mm		
System characteristics	#1	#2	#3
Biomass consumptions (kg/h)	239	222	229
Pyrolysis temperature (°C)	450	450	450
Air (Nm^3/h)	293.60	240.90	249.6
Steam (kg/h)	–	–	10
Air temperature (°C)	25	400	400
Steam temperature (°C)	–	–	400
LHV of produced gas (MJ/Nm^3)	6.39	6.97	6.87
Volume of produced gas[a] (MJ/Nm^3)	575.9	499.30	530.40
Gasification temperature (°C)	950	950	950
Ash (kg/h)	3.47	3.22	3.32
Charcoal[b] (kg/h)	3.27	3.04	3.13
Tar[c] (kg/h)	3.58	3.33	3.43
CHP output			
Electric energy (kW)	329.70	330.40	330.50
Heat energy (kW)	316.36	317.03	317.13
Operating hours per year (h)	7000	7000	7000
Overall recoverable thermal energy (kW)	449.08	423.56	453.59
Air preheating (kW)	–	34.47	34.77
Steam generation (kW)	–	–	35.25
Heat block 1 (kW)	83.75	39.62	32.88
Heat block 2 (kW)	35.63	32.44	33.50
Efficiency of CHP system			
η_{cge} (%)	85.35	86.89	88.20
η_e (%)	28.73	29.69	28.80
η_t (%)	39.13	39.07	39.52
η_{CHP} (%)	67.86	68.76	68.32

a - dry gas, b- 5% of pyrolysis charcoal, c - 5% of pyrolysis tar

5.3 Configuration

The modelled characteristics of gasification plant coupled with IC gas engine is presented in Table 2. The input was considered to be 229 kg/h of dry corn cob with air is preheated up to 400 °C. Other input is set up like in pervious case (temperature of

Table 3. Comparison of gas composition (vol.%) given by the downdraft gasification model, literature review

	Model (in this study) steady state model for downdraft		Da Silva [25] downdraft gasification of corn cob	Senelwa [24] downdraft gasification of P. tomantosa with bark
$T_{gasification}$ (°C)	930	955	930	955
λ	0.25	0.19	0.25	0.19
CO	24.21	24.03	19.00	24.10
CO_2	9.68	9.51	10.30	9.50
H_2	14.84	14.47	15.90	12.90
CH_4	3.00[a]	2.50[a]	3.00	2.50
N_2	48.27	49.22	49.51	51.10
LHV of gas (MJ/Nm^3 dry)	6.22	5.52	5.66	6.11

a - add by user

Table 4. Comparison of gas composition (vol. % db) given by the downdraft gasification model for air first and second CHP configuration

	Air preheating Tair = 400 °C	Air and steam Tair/steam = 400 °C
$T_{gasification}$ (°C)	955	955
λ		
Gas composition (vol.%)[a]		
CO	26.94	25.78
CO_2	8.13	8.95
H_2	17.03	17.61
CH_4	2.5	2.5
N_2	45.41	45.17
LHV of gas (MJ/Nm^3 dry)	6.13	6.05

a - dry basis

pyrolysis and gasification process, percentages of charcoal and tar etc.). The temperature of the producer gas leaving the gasifier is considered to be 500 °C. Table 2 shows the simulated results for produced gas.

The CHP plant provides 330.50 kW of electricity and 453.59 kW of heat for District heating. Heat is recovered from producer gas, exhaust gas and IC gas engine. Heat from producer gas is first used for air and steam preheating. After leaving air preheating unit, temperature of the producer gas is lowered from 500 °C to 292 °C.

Cooled producer gas enters the Heater block 1, where temperature of the gas is lowered to 150 °C. The 32.88 kW of heat is used to heat water from 70 to 85 °C for District heating. Heat for District heating is also recovered from IC gas engine with 317.13 kW and exhaust gas cooling, Heater block 2, with 33.50 kW.

In the configuration studied, the cold gas efficiency (η_{cge}) is 88.20% when considering the energy content of clean producer gas. With reference to the corn cob energy content, the electrical (η_e), thermal (η_t) and overall (η_{CHP}) efficiencies of the CHP plant are 28.80%, 39.52% and 68.32% respectively (Fig. 3).

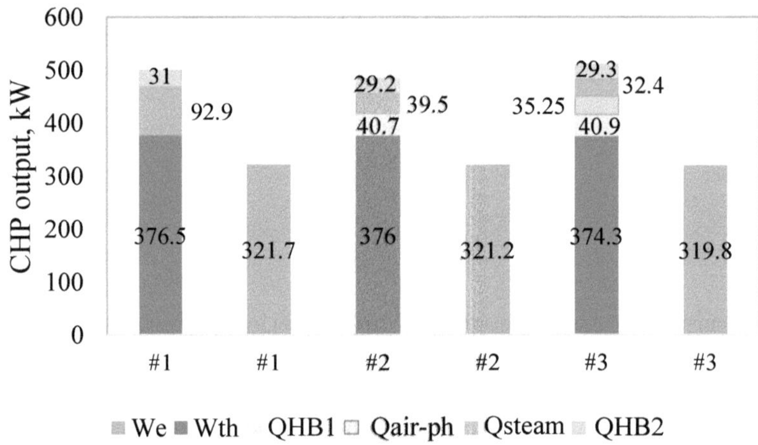

Fig. 3. Comparison of CHP output for different configurations

All three configurations, for adjusted same electrical output of 330 kW, gave similar values for cold gas efficiency, electric and thermal efficiency. However, the overall CHP efficiency is higher for cases when as a gasifying agent is used preheated air and air/steam mixture (around 63%). The use of preheated air and air/steam mixture achieves downsizing of the plant [28, 29]. Downsizing is achieved due to reduced volume of gasifying agent (air or air/steam mixture) which bring the gasifier to the required operating temperature. This also reduced the size of the reactor and gas clean-up system needed [28]. However, according to Puig et al. [5] the air temperature has a significant influence on composition only up to a certain level. It is limited by the effectiveness of the heat-exchange equipment and the operating temperature constraints of the reactor. Mixture of air and steam injected in biomass gasification increases slightly the H_2 content of producer gas. Also, the characteristics of produced gas are better in case of preheated air and steam/air mixture gasification (LHV of approximately 6.9 MJ/m^3). Also, the case where steam/air was used as a gasifying agent at 400 °C, has the highest production of heating for the DH (Fig. 4).

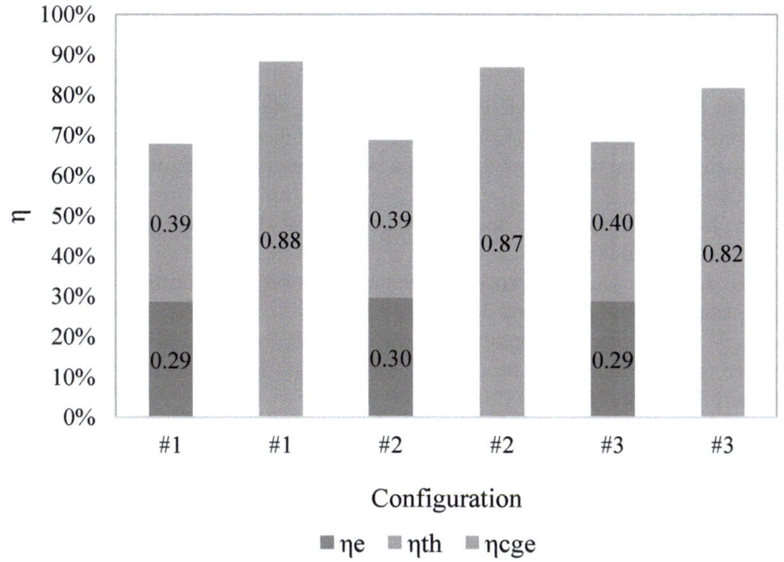

Fig. 4. Comparison of efficiencies for different configurations

6 Conclusions

In the present paper, a scenario analysis of the CHP biomass gasification plant with use of different gasifying agent characteristics is presented.

Model can be used to predict the final producer gas composition and its main characteristics, such as the heating value, for a certain biomass with a defined ultimate composition and moisture, and to predict influence of different gasifying agent characteristics on CHP plant performance. The use of preheated air and air/steam mixture achieves downsizing of the plant [28, 29]. Downsizing is achieved because a smaller volume of gasifying agent (air or air/steam mixture) is needed to bring the gasifier to the required operating temperature, which in turn reduces the size of the reactor and gas clean-up system needed [28]. Mixture of air and steam injected in biomass gasification increases slightly the H_2 content of producer gas. Also, the model has proved to be effective at simulating electricity generation and the composition and gas production. All three configurations generate the same electricity. The third case, with steam/air as gasifying agent at 400 °C, has the highest production of heating for the DH. Nevertheless, all the configurations have similar values for cold gas efficiency (around 82%). The overall CHP efficiency is for 10% higher for cases when as a gasifying agent is used preheated air and air/steam mixture (around 63%). However, all these configurations can be considered to be "high efficiency systems". Choose of suitable configuration depends on user requirements (e.g. production of heating for the DH).

References

1. Gao, N., Li, A.: Modeling and simulation of combined pyrolysis and reduction zone for a downdraft biomass gasifier. Energy Convers. Manag. **49**(12), 3483–3490 (2008)
2. Puig-Arnavat, M., Bruno, J.C., Coronas, A.: Modified thermodynamic equilibrium model for biomass gasification: a study of the influence of operating conditions. Energy Fuels **26**(2), 1385–1394 (2012)
3. Basu, P.: Biomass Gasification and Pyrolysis: Practical Design and Theory, p. 365. Elsevier Inc., Oxford (2010)
4. Francois, J., et al.: Detailed process modeling of a wood gasification combined heat and power plant. Biomass Bioenerg. **51**, 68–82 (2013)
5. Trninić, M.: Modeling and Optimisation of corn cob Pyrolysis, in Faculty of Mechanical Engineering. Department for Process Engineering and Enviromental Protection, Belgrade University Belgrade, Belgrade (2015)
6. Sterner, M.: Bioenergy and Renewable Power Methane in Integrated 100% Renewable Energy Systems. Limiting Global Warming by Transforming Energy Systems. Faculty of Electrical Engineering and Computer Science, University of Kassel, Kassel (2009)
7. Loo, S.V., Koppejan, J.: The Handbook of Biomass Combustion and Co-firing. Earthscan, London (2008)
8. Chinese, D., Meneghetti, A.: Optimisation models for decision support in the development of biomass-based industrial district-heating networks in Italy. Appl. Energy **82**(3), 228–254 (2005)
9. Morris, M., et al.: Status of large-scale biomassgasification and prospects (Chap. 5). In: Knoef, H.A.M. (ed.) Handbook Biomass Gasification, Enschede, Netherlands (2005)
10. Hagos, F.Y., Aziz, A.R.A., Sulaiman, S.A.: Trends of syngas as a fuel in internal combustion engines. Adv. Mech. Eng. **6**, 401587 (2014)
11. Ahmed, T.Y., et al.: Mathematical and computational approaches for design of biomass gasification for hydrogen production: a review. Renew. Sustain. Energy Rev. **16**(4), 2304–2315 (2012)
12. Gómez-Barea, A., Leckner, B.: Modeling of biomass gasification in fluidized bed. Prog. Energy Combust. Sci. **36**(4), 444–509 (2010)
13. Li, C., Suzuki, K.: Tar property, analysis, reforming mechanism and model for biomass gasification: an overview. Renew. Sustain. Energy Rev. **13**(3), 594–604 (2009)
14. Puig-Arnavat, M., Bruno, J.C., Coronas, A.: Review and analysis of biomass gasification models. Renew. Sustain. Energy Rev. **14**(9), 2841–2851 (2010)
15. Ruggiero, M., Manfrida, G.: An equilibrium model for biomass gasification processes. Renew. Energy **16**(1–4), 1106–1109 (1999)
16. Mikulandrić, R., et al.: Artificial neural network modelling approach for a biomass gasification process in fixed bed gasifiers. Energy Convers. Manag. **87**, 1210–1223 (2014)
17. Patuzzi, F., et al.: Small-scale biomass gasification CHP systems: comparative performance assessment and monitoring experiences in South Tyrol (Italy). Energy **112**, 285–293 (2016)
18. Puig-Arnavat, M., Bruno, J.C., Coronas, A.: Modeling of trigeneration configurations based on biomass gasification and comparison of performance. Appl. Energy **114**, 845–856 (2014)
19. Zabaniotou, A., et al.: Bioenergy technology: gasification with internal combustion engine application. Energy Procedia **42**, 745–753 (2013)
20. F-Chart Software: EES-Engineering Equation Solver 2016, Professional Version V 10.066-3D
21. Wang, L., et al.: Is elevated pressure required to achieve a high fixed-carbon yield of charcoal from biomass? Part 1: Round-Robin Results for Three Different Corncob Materials. Energy Fuels **25**(7), 3251–3265 (2011)

22. Trninić, M., et al.: Kinetics of Corncob Pyrolysis. Energy Fuels **26**(4), 2005–2013 (2012)
23. Trninić, M., Jovović, A., Stojiljković, D.: A steady state model of agricultural waste pyrolysis: a mini review. Waste Manag. Res. **34**(9), 851–865 (2016)
24. Senelwa, K.A.: The air gasification of woody biomass from short rotation forests short rotation forests. In: Agricultural Engineering, Massey University, New Zealand (1997)
25. Da Silva, J.N.: Tar Formation in Corncob Gasification, Purdue University, West Lafayette, Indiana, USA (1984)
26. Elliott, M.A., Nebel, G.J., Rounds, F.G.: The composition of exhaust gases from diesel, gasoline and propane powered motor coaches. J. Air Pollut. Control Assoc. **5**(2), 103–108 (1955)
27. GE Jenbacher: Jenbacher gas engines - Jenbacher Type JMS 208 GS-B.L.
28. Doherty, W., Reynolds, A., Kennedy, D.: The effect of air preheating in a biomass CFB gasifier using ASPEN Plussimulation. Biomass Bioenergy **33**(9), 1158–1167 (2009)
29. Sugiyama, S., et al.: Gasification performance of coals using high temperature air. Energy **30** (2), 399–413 (2005)

Determination of the Wing Conveyor Idlers' Axial Loads Using the Finite Element Method

Zarko Miskovic[1], Radivoje Mitrovic[1], Milan Tasic[2(✉)], Marko Tasic[2], and Ján Danko[3]

[1] Faculty of Mechanical Engineering, University of Belgrade, 11000 Belgrade, Serbia
zmiskovic@mas.bg.ac.rs

[2] College of Applied Sciences, Tehnikum Taurunum, Belgrade 11080, Serbia
tasam@sezampro.rs

[3] Faculty of Mechanical Engineering, Institute of Transport Technology and Designing, Slovak University of Technology in Bratislava, Bratislava, Slovakia

Abstract. The impact of the axial load is often completely neglected in the design of the conveyor idlers (rollers) testing machines. The subject of the present research is focused primarily on the conveyor idlers load determination by numerical simulation of the contact between the conveyor belt and idlers. A nonlinear theory of the finite element method is applied, taking into account the effects of large displacements and the contact problems. The basic information required for the presented calculation was the modulus of elasticity of the conveyor belt in the lateral direction. An experimental apparatus in accordance with the DIN 22102 standard was developed and used to determine the load representative value. The adopted approach to the numerical modelling was initially checked by the simulation of the designed experimental testing. A significant match of the results confirmed the applicability of the presented approach to the modelling of the considered problem. The axial load on the wing (side) idlers is generated only during the partial loading of the conveyor. It has considerably high intensity only until the conveyor belt touches the horizontal idler. Applying the gradually increasing load on the conveyor belt in the numerical model and monitoring the vertical distance between the belt and the horizontal conveyor idler, the exact moment of contact was determined. The reaction forces registered in the contact of the belt and the wing idlers are used as the experimental loads in the custom designed conveyor idlers testing machine – where conveyor idlers are tested under the simultaneous action of the radial and axial load.

Keywords: Conveyor idlers · Finite element simulations · Testing procedure Mechanical design

N. Mitrovic et al. (Eds.): CNNTech 2018, LNNS 54, pp. 174–192, 2019.
https://doi.org/10.1007/978-3-319-99620-2_14

1 Introduction

1.1 Belt Conveyors and Their Components

Belt conveyors are most often used for the transportation of the bulk materials on the surface coal mines. The length of the conveyor belt reaches up to 10 km and consists of about 2000 sections. The conveyor belt is usually supported by a steel structure and conveyor idlers organised in garlands. The CAD 3-D model of a conveyor section with basic components is shown in Fig. 1 [1].

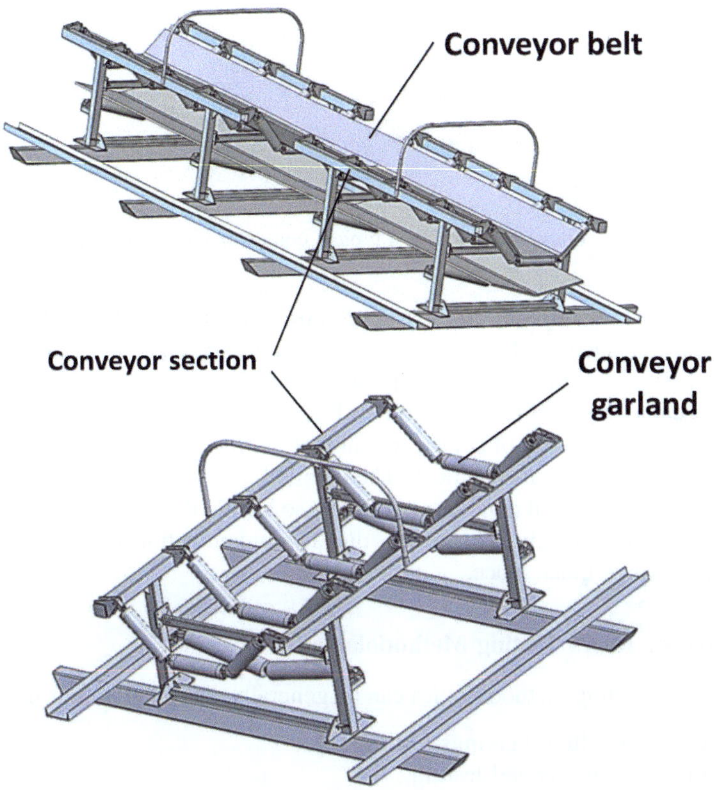

Fig. 1. Conveyor section with the garlands [1].

Conveyor idlers (Fig. 2) generally consist of the shaft, shell, a pair of sealing groups and a pair of rolling bearings [1]. There are more than 80000 conveyor idlers per average open pit coal mine, so any improvement of their reliability leads to a significant improvement of the conveyor systems overall performances.

Rolling bearings are the key components of conveyor idlers. Their failure is the most common reason of the conveyor idler failure as an assembly [2–4]. If bearings are properly installed and maintained, the time of their failure can be accurately predicted

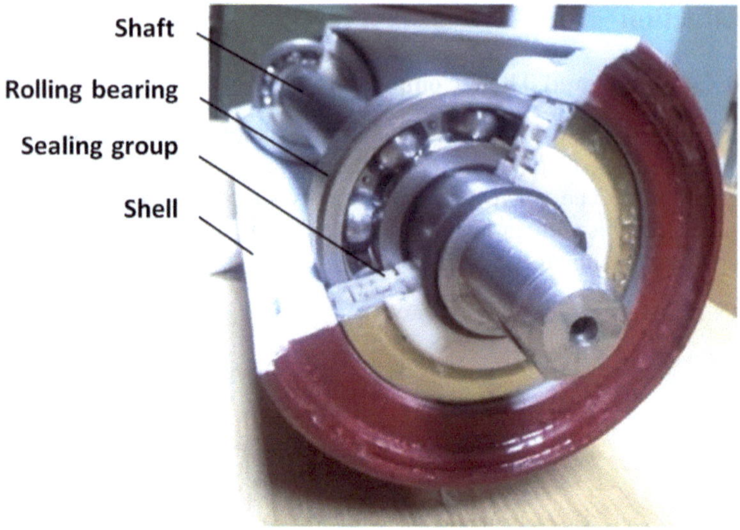

Fig. 2. Component parts of a typical conveyor idler assembly [1].

using the modified equation of their basic life presented in [5] and analysed in detail in numerous papers such as [6, 7]. However, contamination can significantly affect conveyor idlers bearings service life [2–4], which leads to a significant difference between the calculated and measured time of bearing failure. The main reason for this is a very high concentration of solid contaminant particles (coal, dust, sand etc.) in the conveyor idlers working environment. The detailed analysis of conveyor systems has shown that the environment conditions could be so hostile that conveyor idlers bearings eventually become severely damaged – due to contamination particles which have entered the bearings inner space.

1.2 Conveyor Idlers Testing Methodologies

Conveyor idlers testing methodologies can be generally classified in four major groups:

- conveyor idlers rolling bearings testing,
- conveyor idlers accelerated testing,
- sealing groups testing,
- other methodologies (such as friction torque testing or TIR testing).

This paper is focused primarily on the first two groups of the conveyor idlers testing methodologies, while the new methodologies developed by the authors which belong to the other two groups are described in papers [8–11].

Over the last couple of decades, numerous experimental investigations of rolling bearing characteristics were performed (and are still being performed today). The main goal of those investigations is the improvement of rolling bearings reliability, service life and overall efficiency. The examples of relevant research studies are shown in papers [12–15]. However, taking into account the specific environmental conditions of conveyor idlers (a high level of present contamination particles), the most relevant research studies are those which investigate the rolling bearings behaviour in the conditions of abrasive wear.

Bearing that in mind, the new rolling bearings testing methodology has been developed with the aim of determining the correlation between bearings vibration characteristics, operational time and the concentration level of exploitational contamination particles (selected according to the procedure described in [16]) in their grease – for specific working conditions of the open pit coal mine conveyor idlers.

The rolling bearings testing according to the new methodology was performed in following steps:

1. The first pair of sample bearings was greased by 9.3 g of clean grease per sample;
2. The second pair of sample bearings was greased by 9.3 g of grease contaminated by 1 g of prepared particles per sample;
3. The third pair of sample bearings was greased by 9.3 g of grease contaminated by 2 g of prepared particles per sample;
4. The first pair of sample bearings was mounted on the test frame and exposed to the radial load of 4.6 kN and the rotational speed of 1140 RPM;
5. Every half hour, the samples radial vibrations acceleration, expressed as the RMS value, was recorded (Fig. 3);
6. After 10 h of testing, the experiment was finished and sample bearings were dismounted from the test frame;
7. The second pair of sample bearings was mounted on the test frame and exposed to the radial load of 4.6 kN and the rotational speed of 1140 RPM;
8. Every half hour, the sample bearings radial vibrations acceleration, expressed as the RMS value, was recorded;
9. After 10 h of testing, the experiment was finished and sample bearings were dismounted from the test frame;
10. The third pair of sample bearings was mounted on the test frame and exposed to the radial load of 4.6 kN and the rotational speed of 1140 RPM;
11. Every half hour, the sample bearings radial vibrations acceleration, expressed as the RMS value, was recorded;
12. After 10 h of testing, the experiment was finished and sample bearings were dismounted from the test frame.

The results of the performed experimental research are presented in Fig. 4 [17]. The developed mathematical model was generated using the accurate software for statistical analysis and data interpolation with rather high obtained values of the coefficient of multiple determination, which confirmed the validity of the developed equation. At the moment, the presented correlation model is unique and can be successfully used as a basis for further investigation in the relevant scientific area.

During the bearings testing according to the described procedure, the distribution of the inner and outer bearings rings temperatures was also observed. The results of this investigation are shown in Fig. 5 [18] – for the bearings inner ring. The bearings surface temperatures were very consistent. The samples tested in the same conditions

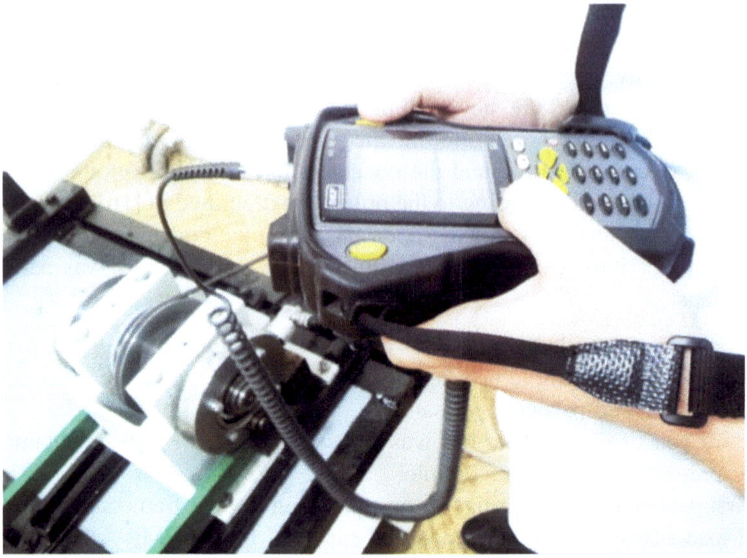

Fig. 3. Sample bearings vibration measurements using the measurement instrumentSKF Microlog Analyzer GX [17].

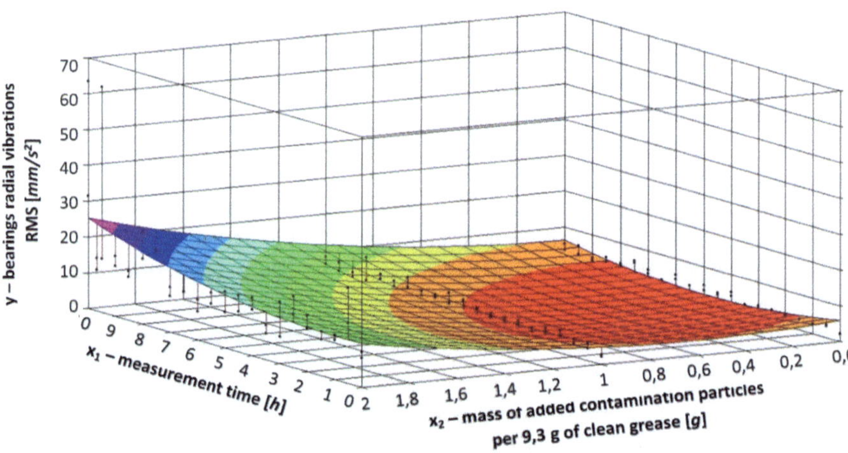

Fig. 4. Graphical presentation of interpolated measured bearings radial vibrations expressed as the RMS [17].

have shown almost identical behaviour and reached thermal balance after 3 to 4 h, with a visible correlation between the mass of solid contamination particles in tested bearings grease and the increment of the bearings temperatures [18]. In all cases, the bearings inner rings temperatures were higher than those of the outer rings. This effect could be explained by the less convenient load distribution due to inner rings smaller diameters.

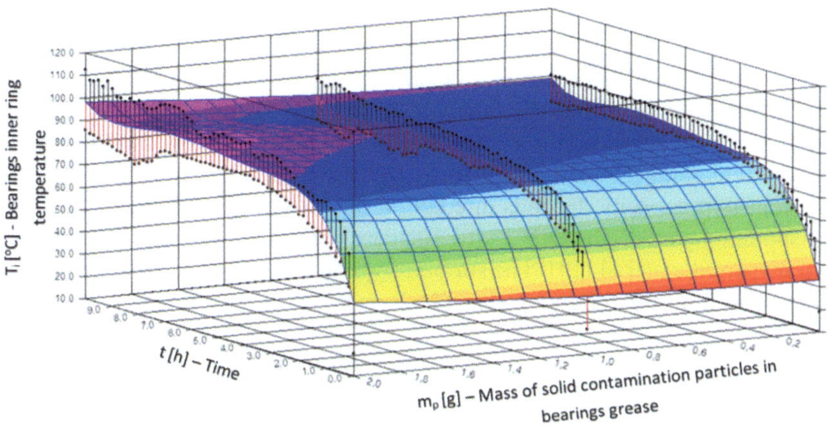

Fig. 5. Graphical presentation of the measured bearings inner rings temperatures, time and mass of solid contamination particles in their grease [18].

The described rolling bearings testing methodology can be successfully used for the testing of the conveyor idlers bearings, with some limitations:

- it simulates operational conditions only for the conveyor idlers located in the middle of the garland – no axial loads are applied, such as one acting on the wing conveyor idlers bearings,
- the mechanical characteristics of other conveyor idler parts (such as the shaft or shell stiffness) and their effect on the rolling bearings performance are neglected – a rolling bearing is tested as a unit separated from the conveyor idler assembly.

In order to overcome the listed limitations, a new conveyor idler testing methodology (including the appropriate testing machine) has been developed and described in Sect. 3. The main advantage of this methodology is the approach where a sample conveyor idler is tested as a full assembly (with all component parts – as it is used in real operational conditions). However, in order to apply the new testing methodology properly, the experimental loading conditions must be precisely defined based on the operational axial and radial loads acting on the garlands wing and the middle conveyor idlers. This objective has been achieved using the appropriate analytical equations, as well as the Finite Element Method.

2 Determination of the Conveyor Idlers Radial and Axial Loads

2.1 Load Distribution in the Transverse Direction

The loaded conveyor belt has a corrugated cross-section which allows it to receive the maximum amount of transported material per unit of belt length. Figure 6 shows the distribution of the idler load in the garland. There is evident correlation between the geometry of the transported material and the load ratio – namely, the change of the depth of the “trough” of the conveyor belt causes the load ratio of the lateral and horizontal idlers to change as well. At the ratio of width b/B = 0.45 and at the inclination angle l = 30°, the horizontal idler takes more than 70% of the total load (e = 0.7).

It is widely accepted that the middle idler carries roughly 50–70% of the static loading due to the conveyor belt and the transported material mass, and the wing idlers accepts the remaining 15–25%.

The transport capacity depends directly on the cross-sectional area of the belt and the transported material. The DIN 22101 standard defines the shape of the cross section of the transported material [19]. The surfaces A_1, A_2 and A_3 in the cross section can be calculated based on the following Eqs. (1, 2, and 3).

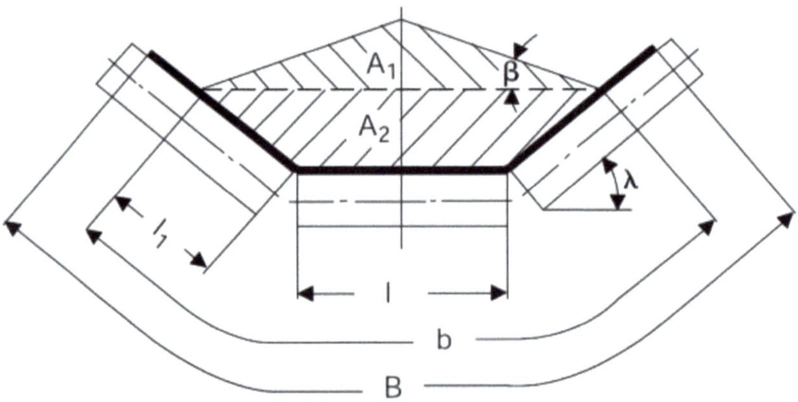

Fig. 6. Distribution of the idler load in the garland [19].

Cross-Sectional Area of the Load Stream:

$$A = A_1 + A_2 \ [m^2] \tag{1}$$

For carrying conveyor idler sets, the cross-sectional area can be calculated as follows:

$$A_1 = 0.25 * \tan\beta * [l + (b - l) * \cos\lambda]^2 \ [m^2] \tag{2}$$

and

$$A_2 = l_1 * \sin\lambda * [l + l_1 * \cos\lambda] \ [m^2] \tag{3}$$

L (m) Length of middle carrying roll
l_1 (m) Loading width of outer rolls
B (m) Usable belt width (load stream width)
λ (°) Troughing Angle
β (°) Surcharge Angle

The transport capacity is the function of the inclination angle of the side idlers. Table 1 gives a short overview of the transport capacity according to the belts manufacturer 'Dunlop'.

Table 1. Theoretical Volume Stream Q_v (m^3/h)

Theoretical Volume Stream Qv (m^3/h)
v=1 [m/s] and inclination= 0^0

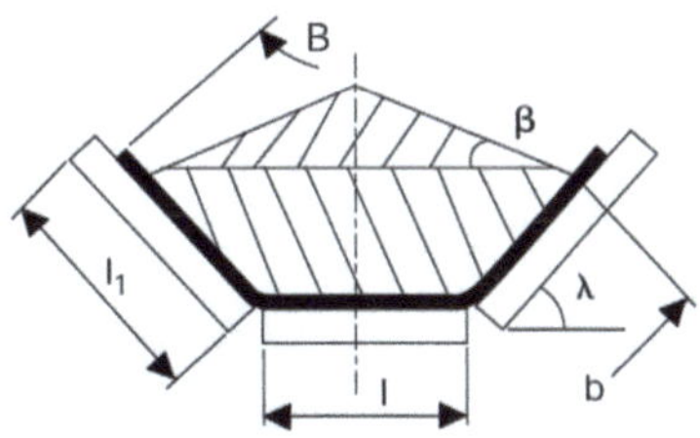

Belt Width B (mm)	Surcharge Angle β	Troughing Angle λ 20°	30°	35°	40°	45°
1200	0	286	401	449	487	516
	10	441	540	577	605	623
	15	521	612	644	666	678
	20	605	687	714	731	737
1400	0	393	552	617	670	711
	10	606	743	795	834	859
	15	717	843	888	919	936
	20	834	947	985	1008	1017

A greater inclination angle gives greater capacity, but at the same time increases the load acting on the wing idlers.

During work (in operational conditions), the conveyor belt has a corrugated shape. The belt material touches all three idlers. In this case, the wing idlers are unloaded in the axial direction. In the case of rigid belts, when they are not sufficiently loaded, the belt will rely only on the wing idlers. As a consequence, the movement of the belt become sunstable and it may be skewed, leading to oscillations in the lateral direction. These oscillations generate a dynamic component of the axial load of the wing conveyor idlers bearings.

The axial load of wing idlers occurs due to:

- the shape of the conveyor path that is not ideal,
- loading materials which are often unsymmetrically distributed,
- weight of the rotational parts of the side conveyor idlers (for a roller Ø159 670 mm, it is 164 N)
- insufficiently loaded conveyor.

Figure 7 shows the contact between the unloaded belt and the wing idlers. The lack of contact with the horizontal conveyor idler is also evident.

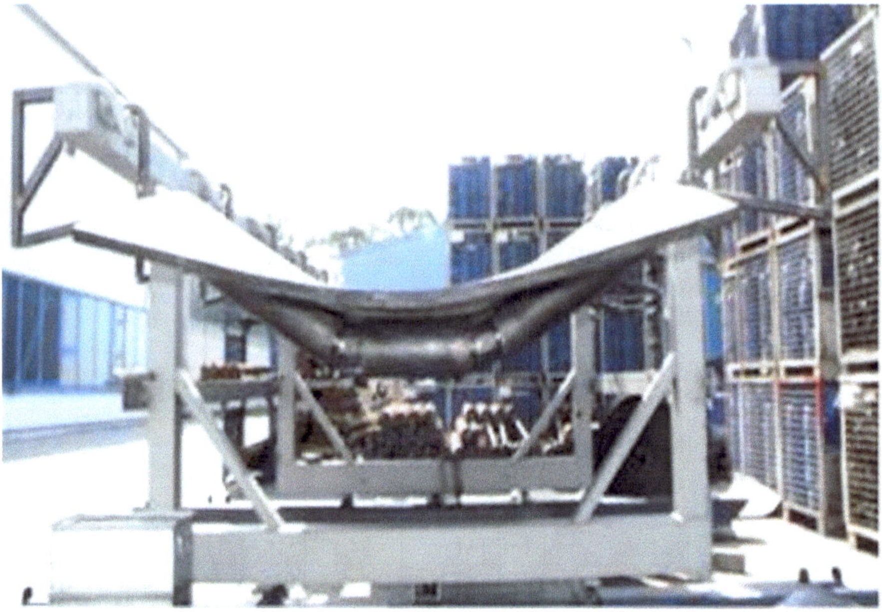

Fig. 7. Contact between unloaded belt and wing conveyor idlers

Figure 8 shows a 3D CAD model of the conveyor idler rotational parts. Their mass, together with the mass of a partially loaded belt, generates the axial load of the conveyor idlers rolling bearings.

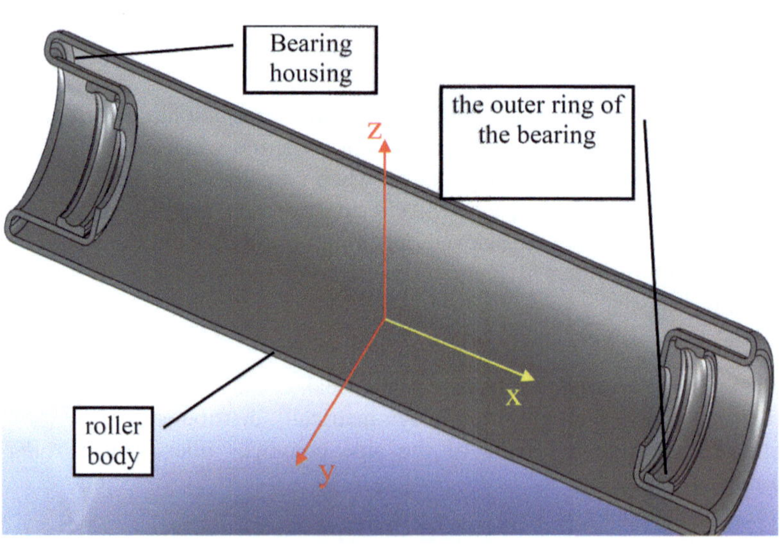

Fig. 8. Contact between the unloaded belt and wing conveyor idlers

2.2 Determination of the Conveyor Belts Modulus of Elasticity in the Lateral Direction

The value of the modulus of elasticity of the belt in the lateral direction is rarely found in the manufacturers catalogues. It is most usually determined experimentally using the installation described in DIN 22102 (Fig. 9) [20].

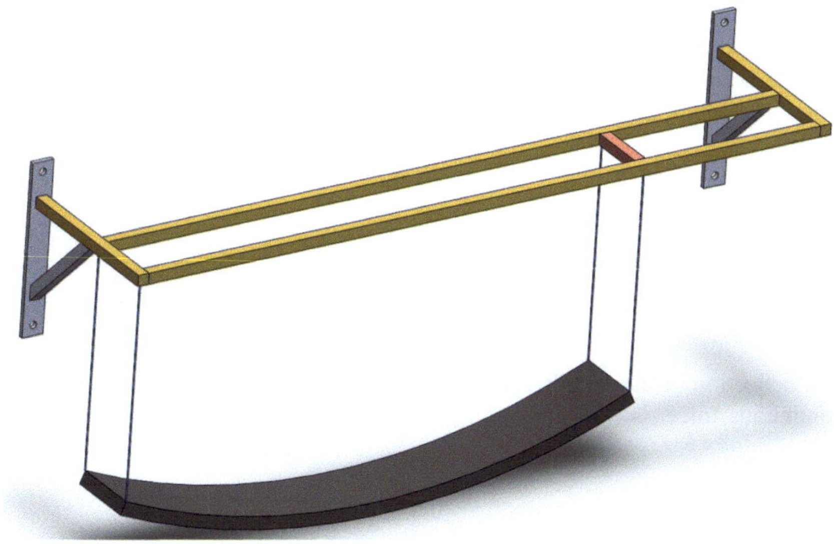

Fig. 9. Determination of the belts lateral modulus of elasticity according to the DIN 22102

The belt manufacturer usually prescribes the allowed ratio between the maximum deflection and the width of the conveyor belt. The data of the 'Dunlop' belts manufacturer are shown in Table 2.

Table 2. The prescribed ratio between the maximum deflection and width of the belt

Inclination angle [°]	20	25	30	35	40	45	50	55	60
Allowed ratio [f/B]	0.08	0.10	0.12	0.14	0.16	0.18	0.20	0.23	0.26

The test sample is a small part of the conveyor belt with a 1600 mm width and the length of 150 mm (Fig. 10).

The determination of the conveyor belt deformation in the transverse direction is performed according to the following steps:

1. The belt specimen is tied at both ends to the hanging wire (500 mm long),
2. Vertical distance is measured at the centre of the sample,
3. Side edges of the specimen should be horizontal,
4. The sample bends only under the influence of the gravity force.

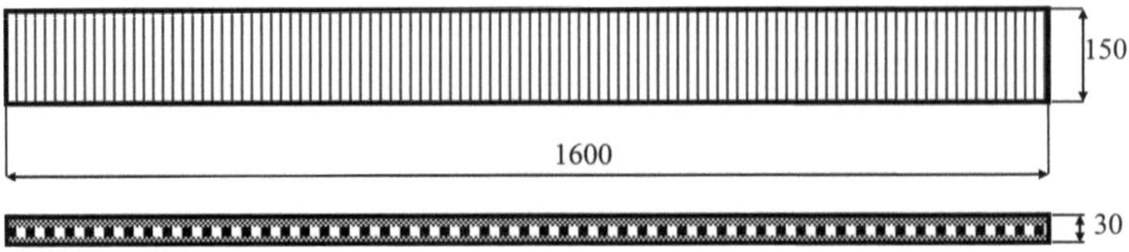

Fig. 10. Geometry of the belt test sample according to the DIN 22102

For the adopted angle of the wing idlers of 20°, the allowed ratio between the maximum deflection and the width of the belt is 0.08:

$$\frac{f}{B} = 0.08 \tag{4}$$

Permissible deflection is calculated according to:

$$f = B * 0.08 = 1600 * 0.08 = 128\,\text{mm}$$

The deflection of the middle-span of the free beam with two supports, loaded with the continuous force, is:

$$f = \frac{5}{384}\frac{q * l^4}{E * I} \tag{5}$$

For the rectangular cross-section of the sample:

$$I = \frac{b * h^3}{12} \tag{6}$$

Where:

q_b specific load due to the weight of the belt sample
l distance between the supports (1600 mm)
E lateral modulus of elasticity
I the moment of the inertia of the strip (section in the cross-section)
B the width of the belt (150 mm)
H thickness of the belt (30 mm)

Specific weight for the 1600 mm belt and transportation of heavy and bulky material is:

$$q_b = 36.0\,\text{kg/m}^2 \tag{7}$$

The specific load per 1 cm of the tested belt length in the lateral direction is:

$$q = 0.529\,N/\mathrm{cm} \tag{8}$$

The modulus of elasticity in the lateral direction of the tested belt specimen is:

$$E = \frac{5}{384}\frac{q * l^4}{f * I} = 105\frac{N}{\mathrm{mm}^2} \tag{9}$$

This value of the modulus of elasticity will be used in the deformation calculations of the belt deformation using the numerical methods.

2.3 Numerical Model of the Test According to the DIN 22102 Standard

Figure 11 shows a finite elements mesh for the numerical model of the test in accordance with the DIN 22102 [21]. The half of the belt model (symmetric) is loaded only by its own weight. The sidelong ends of the belt can slide along an inclined supports surface (immobile wing conveyor idlers).

The calculation was made according to the second-order theory of elasticity (nonlinear, with large displacements).

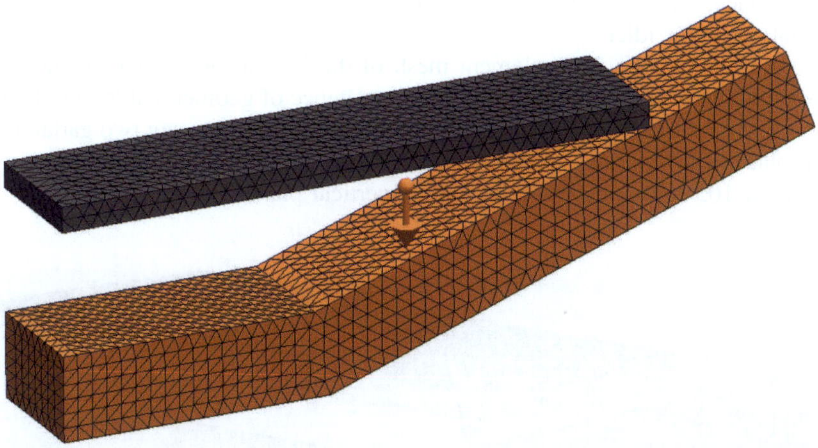

Fig. 11. Numerical model of the test according to the DIN 22102 standard

The field of displacements is shown in Fig. 12. The maximum deflection is in the middle of the support range (in the symmetry plane) and equals 120 mm. The value of the calculated deflection deviates from the value of the deflection angle calculated according to the first-order theory with 128 mm, by 6%. Taking into account the mentioned deviation, it can be concluded that the adopted value of the belt lateral modulus of elasticity is relevant and applicable for the next steps in the finite element method.

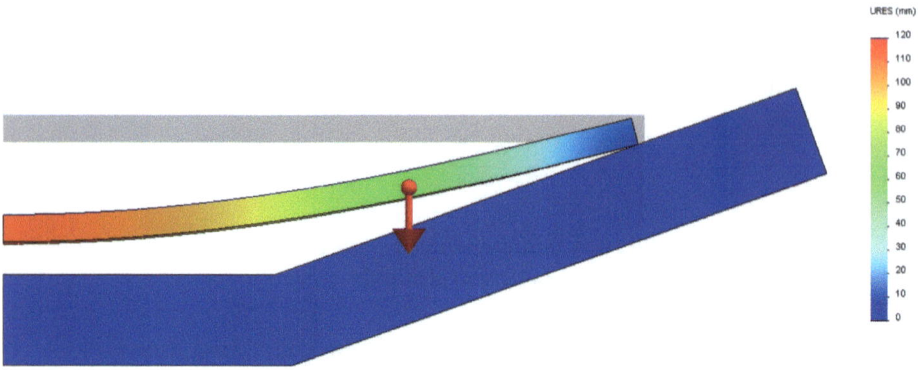

Fig. 12. Displacements of the belt due to its own mass

2.4 The Quarterly Field Model of the Belt Between Garlands

After the selected value for the belt modulus of the elasticity in the lateral direction had been checked and accepted, the calculation of the entire belt deformation between the two garlands was initiated. The goal was to determine the intensity of the load acting on the belt surface, which would lead to the first contact of the belt lower zone with the horizontal conveyor idler.

Figure 13 shows the finite element mesh of the ¼ belt model. The application of one-fourth of the model is possible due to the existence of geometrical longitudinal and transversal symmetry in the support and the load in the field between two garlands. The belts displacement field, due to its own mass, is shown in Fig. 14. The maximum belt deflection of 105 mm appears in the middle vertical plane.

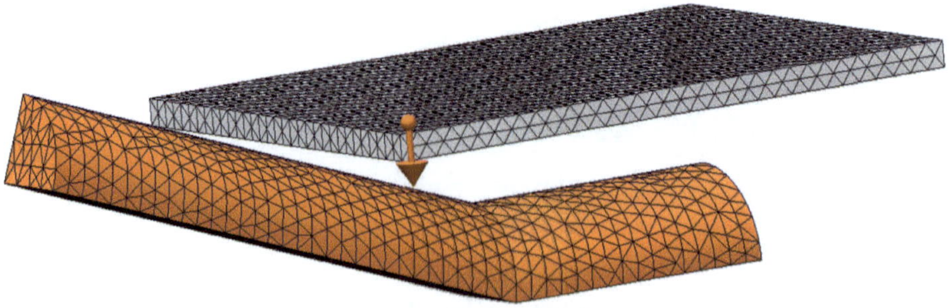

Fig. 13. Finite element mesh of the ¼ belt model

In the left part of the image, a maximal deflection between the garlands (roughly 20 mm) is shown, in the longitudinal direction. As the initial distance between the belt level and the horizontal idler was 146 mm, the rest of the space of 41 mm is available for a deflection due to the load from the transported material.

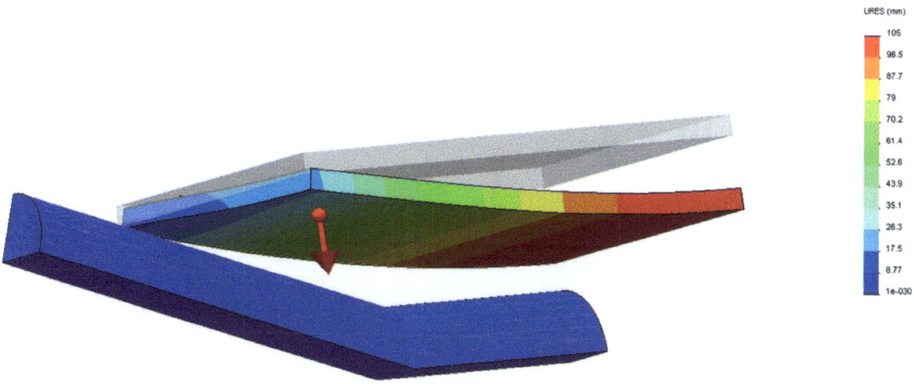

Fig. 14. Displacements field of the belt due to its own weight

The load is applied as uniform pressure on the upper belt surface. The value of pressure is varied (in a series of loading cases) until the contact of the lower belt surface with the cylindrical surface of the horizontal idler is reached. Up until that moment, there exists an axial load on the wing idlers.

The increase of pressure (up to 175 Pa) leads to the contact between the belt lower surface and the horizontal idler, and this value can be accepted as a limit for the estimation of the wing idler axial load (Fig. 15)

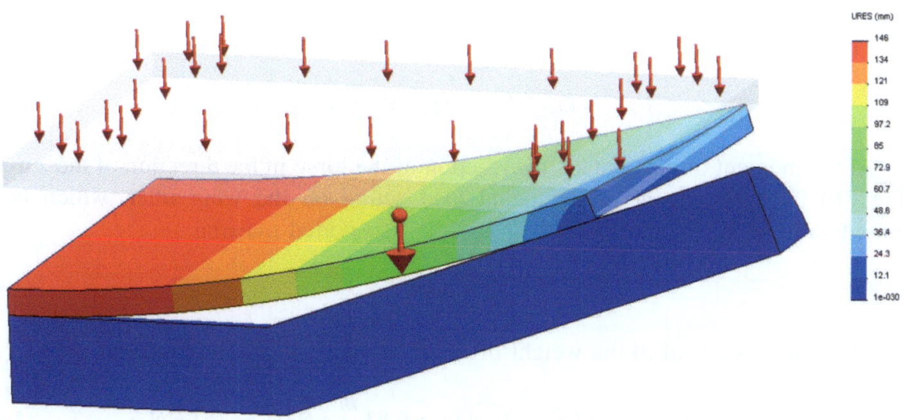

Fig. 15. The moment of contact of the belt and the horizontal conveyor idler (pressure 175 Pa)

2.5 Wing Conveyor Idlers Axial Load Intensity

The results of the reaction force at contact points were exported from the database. The values of these forces are given in Fig. 16.

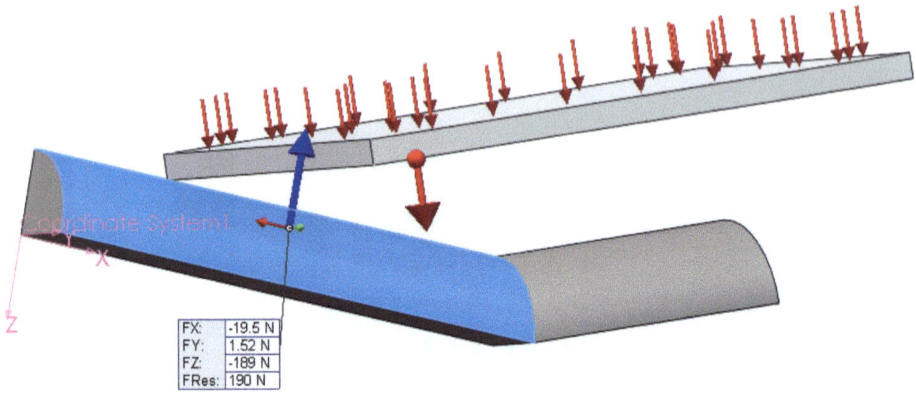

Fig. 16. Reaction forces at contact points of the conveyor idler surface

At the moment of the contact, the force projection on the x axis F_x (Fig. 16) is a component of the wing idler total axial load, and its intensity is:

$$F_x = 19.5N \tag{10}$$

This value is added to the load due to the weight of the wing idler rotary parts:

$$G_x = G_{rot} * sin20^0 = 16.4\,\text{kg} * 9.81\frac{m}{s^2} * 0.342 = 55N \tag{11}$$

The total axial force acting on the wing conveyor idler is:

$$F_{aks} = G_x + F_x = 55 + 19.5 = 74.5N \tag{12}$$

At the moment of the contact, the reaction of the force in the direction of the z axis (Fig. 16) is a component of the radial load acting on the wing idler, which is a consequence of the weight of the belt and the transported material.

$$Fz = 189N \tag{13}$$

This value is added to the weight of the rotary parts of the wing idler:

$$G_z = G_{rot} * cos20^0 = 16.4\,\text{kg} * 9.81\frac{m}{s^2} * 0.94 = 151.2N \tag{14}$$

The total radial force acting on the wing conveyor idlers is:

$$F_{rad} = G_z + F_z = 151.2 + 189 = 340.2N \tag{15}$$

The obtained results were used as a basis for the design of a conveyor idlers testing machine, which tests the conveyor idlers performance under the acting of simultaneous radial and axial forces. The appearance of the reconstructed test frame is shown in Fig. 17.

Fig. 17. The conveyor idler testing machine (*top*) and traces of drive wheels on the surface of the tested conveyor idler *(bottom)*

3 Conveyor Idlers Testing Machine

The developed conveyor idlers testing machine (test bench) in the LIMES laboratory at the Faculty of Mechanical Engineering of the University of Belgrade is intended primarily for radially loaded conveyor idler testing, i.e. not for the testing of wing conveyor idlers. Its basic operational characteristics are described in papers [22–25]. Figure 17 (top) shows the appearance of the conveyor idlers test bench.

The bottom part of Fig. 17 shows the traces of drive wheels on the surface of the shell of the tested conveyor idler. The air pressure in the pneumatic cylinders at both ends defines the radial loading of the idler. A set of six wheels generates a testing radial force and it acts on the entire length of the conveyor idler body.

The intensities of the axial force on the wing conveyor idlers are small in comparison with the radial force. Only one drive wheel is needed to generate them. Simultaneous acting of the loads radial and axial components is achieved using an inclined position of the tested conveyor idler. The appearance of the reconstructed conveyor idler testing machine is shown in Fig. 18.

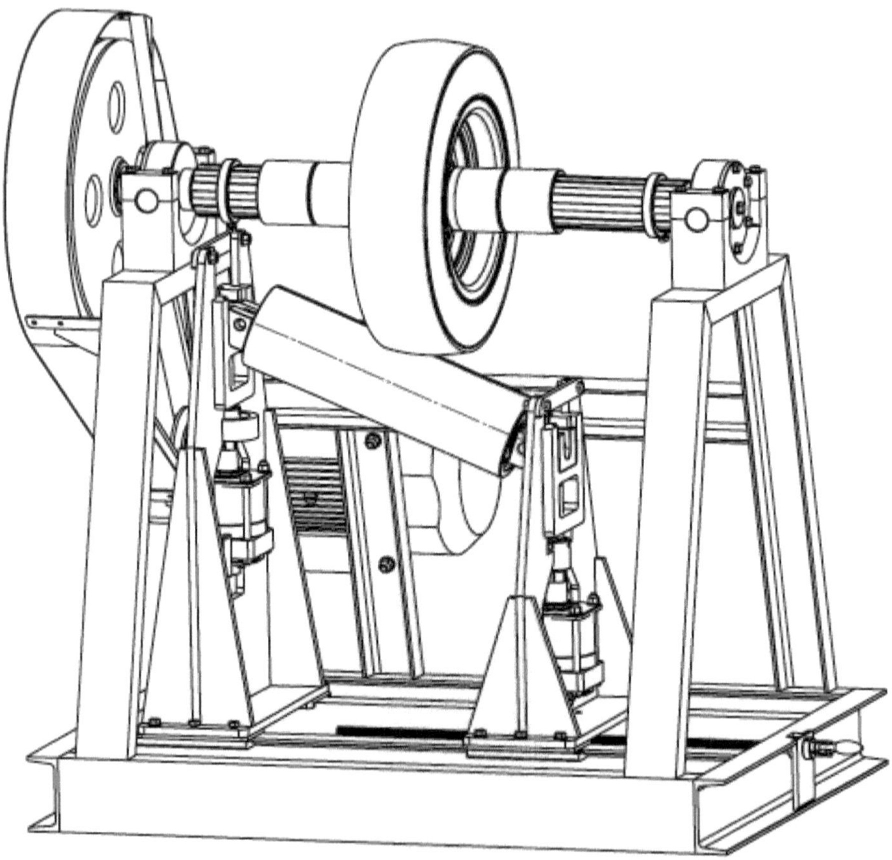

Fig. 18. Conveyor idler testing machine in a position for wing conveyor idlers testing

4 Conclusion

There are very little experimental data available on the axial force intensities acting on the wing conveyor idlers. In the absence of experimental data, it is possible to use numerical simulations in order to determine the axial force with satisfying accuracy. The simulation using the nonlinear Finite Element Method software (FEM) was used very successfully.

The results of the presented analysis confirmed the significance of the axial load acting on the wing conveyor idlers – especially when the conveyor belt is unloaded (or loaded with a small mass of the transported material). As the conveyor system very often operates in the described conditions, it is really necessary to provide the possibility to conduct conveyor idlers testing which also simulates the influence of dynamic axial loads. This objective has been fully achieved now on the basis of the results of the presented calculations and analysis.

Acknowledgement. The authors would like to express their sincere gratitude to the Ministry of Education, Science and Technological Development of Republic of Serbia for the support, as well as for the equipment used, acquired within the national project TR35029.

References

1. Mišković, Ž.: Exploitational contamination particles concentration influence on rolling bearing's operational characteristics. Ph.D. thesis, University of Belgrade, Serbia (2017)
2. Madden, W.F.: An investigation into idler seals and the effect on rim drag. In: The Key Center for Bulk Solids and Particular Technologies, Beltcon (1997). The Web Version: http://www.saimh.co.za/beltcon/beltcon9/paper918.html. Last accessed 20 June 2018
3. Stewart-Lord, M.: Rolling bearing characteristics required for maximizing the life of conveyor idler rolls. In: Beltcon (1991). The Web Version: http://www.saimh.co.za/beltcon/beltcon6/paper66.html. Accessed 20 June 2018
4. Inpro/Seal Company Rock Island, IL. http://www.mtonline.com/january2007/-problem-solvers-conveyor-idler-bearing-isolator-increases-reliability-and-safety-while-reducing-downtime-and-power-consumption. Accessed 20 June 2018
5. Palmgren, A., Lundberg, G.: Dynamic capacity of rolling bearings. In: Acta Polytechnica, 1st edn, pp. 3–7 (1947)
6. Mitrović, R.: Study on influence of design and tribology parameters on rolling bearings working performances during high speed rotations. Ph.D. thesis, University of Belgrade, Serbia (1992)
7. Mitrović, R., Ristivojević, M., Lazović, T.: Different approaches to the rolling bearings' service life assessment. In: Proceedings from the IX SEVER Symposium on Mechanical Transmission Systems, Subotica, Serbia, pp. 17–22 (2003)
8. Mitrović, R., Soldat, N., Mišković, Ž., Matić, N.: Some experiences in laboratory testing of bearings of transport idlers on belt conveyor. In: Proceedings of the 11th International Conference DEMI 2013, Faculty of Mechanical Engineering, Banja Luka, Bosnia and Herzegovina, pp. 207–212 (2013). ISBN 978-99938-39-45-3
9. Mitrović, R., Mišković, Ž., Ivanović, G., Tasić, M., Stamenić, Z.: Development of experimental methodology for conveyor idler's sealing group testing. In: Proceedings of the 2nd International Scientific Conference COMETa 2014, Faculty of Mechanical Engineering, University of East Sarajevo, East Sarajevo—Jahorina, Bosnia and Herzegovina, pp. 497–504 (2014). ISBN 978-99976-623-1-6
10. Mitrović, R., Mišković, Ž., Tasić, M., Stamenić, Z.: Conveyor idler's turning resistance testing methodology. In: Machine Design, Faculty of Technical Sciences, University of Novi Sad, Novi Sad, Serbia, vol. 6, no. 4, pp. 107–112 (2014). ISSN 1821-1259
11. Mitrović, R., Mišković, Ž., Stamenić, Z., Bakić, G., Đukić, M., Rajičić, B.: The new experimental testing methodology for conveyor idler's fits control. In: Proceedings on CD, 14th International Conference on Fracture—ICF 2014, Rhodos, Greece, pp. 1–5 (2017)

12. Lazović, T., Mitrović, R., Ristivojević, M.: Influence of internal radial clearance on the ball bearing service life. J. Balkan Tribol. Assoc. Sci. Bulg. Commun. **16**(1), 1–8 (2010). ISSN 1310-4772
13. Atanasovska, I., Mitrovic, R., Momcilovic, D., Subic, A.: Analysis of the nominal load effects on gear load capacity using the finite element method. Proc. Inst. Mech. Eng. Part C: J. Mech. Eng. Sci. JMES2508 **224**(11), 2539–2548 (2010). https://doi.org/10.1243/09544062. ISSN 0954-4062
14. Subić, A., Mitrović, R.: A general approach to rolling bearing analysis. FME Trans. **2**, 7–10 (1992)
15. Bakić, G., Đukić, M., Lazović, T., et al.: New methodology for monitoring and prevention of rotating parts failures. FME Trans. **35**, 195–200 (2007)
16. Mitrović, R., Mišković, Ž., Maksimović, V., Jovanović, D., Ivanović, G., Stamenić, Z., Tasić, M.: Analysis and characterization of coal mine conveyor idlers contamination particles. In: Sixteenth Annual Conference YUCOMAT 2014, Materials Research Society of Serbia, Book of Abstracts, Herceg Novi, Montenegro, p. 96 (2014)
17. Mišković, Ž., Mitrović, R., Maksimović, V., Milivojević, A.: Analysis and prediction of vibrations of ball bearings contaminated by open pit coal mine debris particles. In: Technical Gazzete (Tehnicki Vjesnik), Josip Juraj Strossmayer University of Osijek, Slavonski Brod, Croatia, vol. 24, no. 6 (2017). https://doi.org/10.17559/tv-20151203140843. Print: ISSN 1330-3651, Online: ISSN 1848-6339
18. Miškovic, Ž., Mitrović, R., Stamenić, Z.: Analysis of grease contamination influence on the internal radial clearance of ball bearings by thermographic inspection. In: Thermal Science, Institut za nuklearne nauke "Vinča, Belgrade, Serbia, vol. 20, no. 1, pp. 255–265 (2016). ISSN 0354-9836
19. DIN 22101: Stetigförderer, Gurtförderer für Schüttgüter, Grundlagen für die Berechnung und Auslegung. Deutsches Institut für Normung e.V. February (1982)
20. DIN 22102-1: Textil-Fördergurte für Schüttgüter - Teil 1: Maße, Anforderungen, Kennzeichnung. Deutsches Institut für Normung (2014)
21. Rieg, F., Hackenschmidt, R., Alber-Laukant, B.: Finite Elemente Analyse für Ingenieure. Carl Hanser Verlag, München (2014). ISBN 978-3-446-44283-2
22. Mitrović, R., Tasić, M., Mišković, Ž., Stamenić, Z., Jovanović, D.: Data acquisition and automatisation of a conveyor idler test stand. In: Advanced Materials Research, Trans Tech Publications, Zurich-Durnten, Switzerland, vol. 633, pp. 277–289 (2013). https://doi.org/10.4028/www.scientific.net/AMR.633.277. ISBN 1022-6680/978-3-03785-585-0
23. Mitrović, R., Mišković, Ž., Tasić, M., Stamenić, Z., Soldat, N., Matić, N.: Conveyor idlers testing machine. In: The 29th Danubia-Adria-Symposium on Advances of Experimental Mechanics, Serbian Society of Mechanics, Proceedings, Belgrade, Serbia, pp. 278–281 (2012). ISBN 978-86-7083-762-1
24. Mitrović, R., Stamenić, Z., Mišković, Ž., Tasić, M., Jovanović, D.: Installation for carrier roller idlers of belt conveyors testing on the open pit mining. In: Proceedings of the 7th International Scientific Conference—Research and development of Mechanical Elements and Systems—IRMES 2011, Mechanical Engineering Faculty, University of Nis, Zlatibor, Serbia, pp. 383–388 (2011). ISBN 978-86-6055-012-7
25. Mitrović, R., Stamenić, Z., Mišković, Ž., Tasić, M..: Laboratory installation for belt conveyors idlers testing on servohydraulic testing machine ZWICK HB-250. In: The 7th International Scientific Conference—Research and Development of Mechanical Elements and Systems—IRMES 2011, Mechanical Engineering Faculty, University of Nis, Zlatibor, Serbia, pp. 371–376 (2011). ISBN 978-86-6055-012-7

Mathematical Modelling Approach of WntSignalling PATHWAY Analyse in Alzheimer Disease

Natasa Kablar(✉)

Lola Institute, 11030 Belgrade, Serbia
natasa2017.kablar@gmail.com

Abstract. Alzheimer disease is followed by accumulation of amyloid plaques and neurofibrillary tangles in neural cells of brain, what leads to toxicity and cell dead. These physical brain impairment or damage is followed with intellectual and cognition fall, and loss of capabilities. Beneath these impairments is production of amyloid beta protein and tau protein, that is interfered with signalling pathways that provide important functioning to organism, cells, or central nervous system, like cell proliferation, differentiation, adhesion, survival, and apoptosis, as an examples of functions that are provided with Wnt signaling pathway. Other pathways found to be included in pathology of Alzheimer's disease are AMPK, mTOR, Sirtuin1, and PCB-1. They crosstalk with other molecular mechanisms, biological functions, and cell signaling in normal cell functioning and in disease. However, disease appears in case of abnormalities, irregularities or dysfunctions. Causes of Alzheimer disease are environmental, biological, or genetics factors. It can be also triggered earlier if other diseases are present like pneumonia, diabetes, injuries and strokes, HIV, or other specific diseases. It is interesting to examine connections between these abnormalities, irregularities, and dysfunctions with physical causes of disease, in case of Alzheimer disease formation of amyloid plaques and neurofibrillary tangles. Further, it is important to discover what causes dysfunctions of signaling pathways, or how they fight against risks and factors that cause the disease. In further research we will explore complexity and cross talk between the pathways and connection with diseases. In this paper we are particularly interested for *Wnt signaling pathway* since it has important cell functions, and cause transcription of genes that provide normal functioning and also have protective and neuroprotective effect in preventing and fighting off risk factors of disease. Wnt signaling is also found in other human diseases such as cancer, metabolic diseases, coronary disease, diabetes and obesity, etc. Our aim is to set framework for research examination of Alzheimer disease via understanding molecular mechanisms, biochemical reactions and mathematical modelling, and to perform dynamical analyze with simulation results, stability and bifurcation tests, in order to get better understanding of signaling pathways and connection with diseases.

Keywords: Alzheimer disease · Cross talk of signalling pathways
Wnt signaling pathway · Molecular mechanism · Biochemical reactions
Mathematical model

N. Mitrovic et al. (Eds.): CNNTech 2018, LNNS 54, pp. 193–207, 2019.
https://doi.org/10.1007/978-3-319-99620-2_15

1 Introduction

Alzheimer disease is neurodegenerative disorder which appears in people 65 years old or older, but early onset can appear after 40-ies. It prevents people from ordinary activities, social relationships, and it cause sharper intellectual fall. It is named after Alois Alzheimer who was the first to study Alzheimer disease. Disease can be confirmed only after autopsy, and in living person it is only diagnosed as "probable Alzheimer disease". It is similar with dementia or senile dementia, but pathological and functional signs and brain damage in Alzheimer disease is much harder, [1].

Alzheimer disease is followed with neuropathology which signs are accumulated amyloid plaques on extra cellular surface of brain cell and neurofibrille tangles inside brain cell that prevents brain and neuronal cells form normal functioning. Amyloid plaques and neurofibrille tangles presents specific physical abnormalities in brain.

Alzheimer disease is progressive, what means that symptoms are worsen when disease advance. The earliest symptoms of Alzheimer disease are mild memory loss, forgotten last conversation data, or what year is. It can be present some form of disorientation, problems with routine tasks, changes in personality, or in judgment. As disease progresses, symptoms get worse in everyday life, so 24 h help is needed. It can be present anxiety, suspiciousness and agitation, wondering, difficulties in recognizing members from close family or friends, and dream disturbances. In more advanced form, patient loss ability to talk, have weight and appetite loss. It is not directly fatal, but it can leave patients with disease over decades. It can also cause infections and other diseases like pneumonia which can be ultimate cause of death. Often, people are exhausted and die earlier, but patients who are normally healthy can survive over decades.

Alzheimer disease is characterised with anatomical changes that forms amyloid plaques and neurofibrille tangles that can be perplexed. Tangles and plaques prevents normal functioning and different functions of neurons and neuronal cells in brain, what causes general slowness and loss of capabilities. Amyloid plaques are formed, stacked, and accumulated on extracellular side of neuronal cell in brain. Amyloid is protein that is normally present in whole body. In AD it is unequally divided creating the substance called beta amyloid that is toxic for nerve cell. As amyloid plaques are accumulated brain cells starts to die. Neurofibrille tangles which can be perplexed are second anatomical sign of Alzheimer disease. Normally, each cell in brain contains long fibres made from proteins which serves as scaffold and supports brain cell in right form. It also helps in transport of food materials and nutrients in cell. In AD these fibres starts curling and clewing. Brain cell lose its form and becomes incapable to transfer food and nutrients in right way, what cause cell to eventually die. As plaques and tangles are accumulated in brain, wide cells dying in brain is present.

There are reported several possible causes in development of AD. It seems that Alzheimer disease is caused environmentally and that biological and genetic factors are included. If specific other diseases are present the risk is wider or higher. Scientists have discovered that many people with this form of the disease have a *specific genes abnormality*: mutation in genes located on chromosomes 1, 14, and 21. Furthermore, chromosome 19 contains a gene called APOE which helps in carrying cholesterol in the blood and in recovering nerves after injury. It is observed that people with apoE4 gene have increased risk of developing AD. In addition to genetic factors, *manybiological factors* have been implicated in AD: for example, free radicals which are formed when the body metabolizes oxygen. Free radicals serve important functions - such as helping the immune system to fight disease. However, too many free radicals cause problems. It is observed that brain cells in AD produce the mutated form of amyloid protein and produce more free radicals. However, free radicals can also enforce beta amyloid protein production. Third, *several environmental factors* contribute to AD: aluminium as common contaminant in drinking water, since it is observed that both the plaques and tangles in AD contain aluminium. Important environmental factors are also zinc, smoking, high exposure to paint solvents, exposure to electromagnetic fields and power lines.

Alzheimer disease can also be developed in patients who already have some disease, like injuries, pneumonia, diabetes, strokes, viral infections, HIV, or other. It is important to conduct research to find relationships with these and other diseases. Environmental factors can trigger Alzheimer disease or cause symptoms to appear earlier. Currently, much more research is needed to identify other triggering factors, and to learn what can be done to prevent it.

In order to treat AD drugs are developed and clinically tested. It is needed 10–20 years of research of new drug, prior clinical tests and medical use. The only allowable drugs today are so called cholinergic drugs as tacrine, donepezil, rivastigmine, and galantimine. So far, these drugs had limited success in treating AD. Tacrine causes side effects and is rarely prescribed. Aside of cholinergic drugs, researchers looks for other kinds of drugs that can influence other chemicals in brain, or which interfere with forming plaques or tangles in AD, or enforce brain activity for producing new neurons in order to substitute dead neuronal cells. Also, dietal supplements can help, prevent or slow AD. Damage in brain caused by free radicals can be treated with antioxidants that can be found in Vitamin C, E, beta carotene which is associated to Vitamin A. Other antioxidants found are ginko biloba and phosphatidylserinme. However, supplements with high dosage of antioxidants can cause specific side effects. Safer way in consuming antioxidants is by diet, taking fruits and vegetables, brown rise, integral grain, meat, eggs, and milky food. It can be also used anti flammation drugs like aspirin or pain relief drugs.

It is unclear so far why plaques and tangles start to form in AD brain. Many researchers try to answer these questions and try to develop ways of preventing or healing this neurodegenerative disease. Last researches discovered vaccine that promise preventing and treating of AD since it has possibility of solubility of plaques. However, it is still not found vaccine which will dissolution or solve problems with tangles which is considered to have complicated role in AD, [1].

Research of AD is actual today, but it is considered that will be needed years or decades to solve this problem and to find drug for people altered or ill from AD. In order to find relationship of formation of amyloid plaques and neurofibrille tangles in Alzheimer disease research is conducted to find molecular mechanisms or signalling pathways that cause or influence abnormal or irregular functioning or signalling, what consequently cause formation of these neuronal physical abnormalities, alteration in function, or different cell function, and what lead to development of Alzheimer disease.

For that reason we present findings of signalling pathways that are connected with Alzheimer disease. They cross talk with other molecular mechanisms, signalling pathways, or cause gene transcription that might lead to genetic mutation and alter or change cells functions or fate. Irregularity or dysfunction of signalling pathways can cause AD related problems. Signalling pathways can also act neuroprotectively to neural cells in brain, or heal them after injury or take off the risks if they are attacked.

2 Roles of Signalling Pathways and Crosstalk of Complex Network of Signalling Pathway

Alzheimer disease is characterized with progressive loss of cholinergic neurons that lead to severe behavioural, motor and cognitive impairments. Extracellular amyloid beta plaques and neurofibrillary tangles containing hyper phosphorylated tau are frequently present in brain of patients with AD. Energy failure in neurons and brain in AD is also hallmark of this disease. Energy demands are prerequisite for neuronal communication. It is a question what causes these impairments, energy failure and accumulation of amyloid plaques and neurofibrillary tangles?

In order to answer this question there are attempts to discover metabolic pathways that might lead to pathogenesis of AD. So far, several signalling pathways are discovered, with attempts to discover mechanism behind these pathways: AMPK pathway, mTOR pathway, Sirtuin1 pathway, PGC-1 pathway, and Wnt pathway (name derived after *wingless* gene in *Drosophilia*). Short names are for biomolecules that play important role in cell, [2, 3]. They are able to *modulate several pathological events* in AD. These include *reduction of amyloid beta aggregation and inflammation*, *regulation of mitochondrial dynamics*, and *increased availability of neuronal energy* [1]. They can provide new therapeutic to slow down or prevent development of AD. These pathways normally increase transcription of genes that are important for normal functioning. For example, in case of dis-functioning caused by oxidative or inflammation insult, genes are transcripted in mitochondria to stabilize the functioning. It is important to reach greater understanding of the molecular basis of these pathways and ways how they interact within cell in order to slow down or attenuate metabolic deficits observed in AD.

In this paper we will consider Wntsignalling pathway, in particular canonical Wnt/Beta catenin signalling pathway in order to show in which cellular processes it is included, to show main components of the pathway, and to set up biochemical and mathematical model of this signalling pathway in order to obtain simulation results and better insight into biological mechanism. It is question of further research to see what genes are transcribed and with what cell function, how it relates to proteins, genes, or other compounds in developing amyloid plaques and neurofibrillary tangles in developing Alzheimer disease, and in order to find targets for potential drug development.

Molecular mechanism is still unclear with several genes or proteins associated that leads to Alzheimer disease. The aim of this paper is to present clues how to investigate pathways. For modelling we use computational approach. It is presented biochemical and mathematical framework for dynamical analysing of signalling pathways. More complex problems will be focus of further research, since thorough investigation of the results from literature is needed.

Wntsignalling pathway is involved in several key cellular processes associated with cell proliferation, differentiation, adhesion, survival, and apoptosis in several catabolic and anabolic cells, including neuronal and glial cells which are key residents of Central Nervous System. Wnt is family of secreted cysteine rich glycosylated protein that are named after Drosophilae protein wingless and mouse protein int-1. So far 19 of 24 Wnt genes are identified in humans, with 80 Wnt target genes in humans, mice, *drosophilae*, *xenopus*, *zebrafish* [4, 5].

Wntsignalling pathway is induced with Wnt that binds to frizzled transmembrane receptors located on the cell surface leading to the induction of at least three downstream signalling pathways, [5]:

1. The first is known as *canonical Wnt pathway* which regulates gene transcription through beta catenin, also called Wnt/beta catenin.
2. The second is *non canonical Wnt pathway* that modulates Ca+ release, also called *Wnt/Ca2+ signalling pathway*.
3. Third one is *Wnt cell polarity signalling pathway*, in which Jun N-terminal kinase (JNK) plays a key role and is also called *Wnt/PCP-JNK signalling pathway*.

In this paper we are interested in canonical Wntsignalling pathway.

3 WntSignaling Pathway

The *canonical Wnt/β-catenin signalling pathway* is important for essential cellular functions where it plays key role and is implicated in many diseases. It has role in cell proliferation and specification during development, in cell maintenance and in wound repair [1]. Dysfunction of *Wntsignalling pathway* leads to many diseases or pathological conditions that include neurodegenerative diseases. The pathway dynamics is still not well understood and further clarifications are needed.

It is known that Alzheimer disease is followed with accumulation of amyloid plaques and neurofibrillary tangles, but the question is *how are they related to Wntsignalling pathway (and other pathways) and how they are caused in AD in general*? It is known that Wntsignalling pathway components are altered in AD. Also, it is known that beta catenin levels in AD are *reduced* what consequently causes inherited mutations in presenilin gene, as an example [2].

4 Description of the Mechanism

Here we present *canonical Wntsignalling pathway* with simple mechanism and key components that are identified to be important for which we set biochemical reactions and derive mathematical model.

Molecular mechanism [5] on which the model is based is described as: Wnt binds to cell-surface receptors that transduce a signal via a multistep process involving Dishevelled (Dsh) to the so-called destruction complex (Y), which contains forms of

Axin, adenomatous polyposis coli (APC), and glycogen synthase kinase (GSK-3). In the absence of a Wnt signal, the Y actively degrades β-catenin by phosphorylation and degradation. Following Wnt stimulation, degradation of β-catenin is inhibited through phosphorylation of Y members, leading to accumulation in the cytoplasm of free β-catenin, which is able to translocate to the nucleus where it can form a complex with T-cell factor (TCF) and lymphoid-enhancing factor proteins and influence the transcription of target genes, [6].

The presented model is focused on activation, sequestration, biochemical interactions, phosphorilation, and degradation of key components of main components Dsh (Disheveled), Y (destruction complex), Beta-catenin, and P (phosphatases), TF (transcription factor)in signalling pathway, and localization of these components in cytoplasm and nucleus, and their transport from cytoplasm-to-nucleus, and from nucleus-to cytoplasm.

5 Biochemical Reactions

From the molecular mechanism described above, set of biochemical reactions is obtained. In order to present the model, we first need table of biochemical species which are used in biochemical model, with corresponding symbols and mathematical variables (Table 1).

Table 1. Biochemical species with symbols and mathematical variables

Species	Symbol	Variable
Dishevelled inactive in cytoplasm	D_i	x_1
Dishevelled active in cytoplasm	D_a	x_2
Dishevelled active in nucleus	D_{an}	x_3
Destruction complex inactive in cytoplasm	Y_i $(APC/Axin/GSK3\beta)_i$	x_4
Destruction complex inactive in cytoplasm	Y_a $(APC/Axin/GSK3\beta)_a$	x_5
Destruction complex inactive in nucleus	Y_{ni} $(APC/Axin/GSK3\beta)_{in}$	x_6
Destruction complex inactive in nucleus	Y_{na} $(APC/Axin/GSK3\beta)_{an}$	x_7
Phosphatase in cytoplasm	P	x_8
Phosphatase in nucleus	P_n	x_9
Beta catenoin in cytoplasm	β-catenin	x_{10}
Beta catenoin in nucleus	β-$catenin_n$	x_{11}
Gene transcription factor in nucleus	TF_n	x_{12}
Transcription complex in nucleus: Beta catenin:TCF	IC_{xt}	x_{13}
Intermediate complex in cytoplasm: Beta catenin: Dishevelled	IC_{yd}	x_4

(*continued*)

Table 1. (*continued*)

Species	Symbol	Variable
Intermediate complex destruction complex in nucleus: Beta catenin: Dishevelled	IC_{xdn}	x_{15}
Intermediate complex in cytoplasm: Destruction complex: Phosphatase	IC_{yp}	x_{16}
Intermediate complex in nucleus: Destruction complex: Phosphatase	IC_{ypn}	x_{17}
Intermediate complex in cytoplasm: Beta catenin: Destruction complex	IC_{xy}	x_{18}
Intermediate complex in nucleus: Beta catenin: Destruction complex	IC_{xyn}	x_{19}

Next, we give set of biochemical reactions, with forward and backward reactions rate for each of the reactions.

Here we state few proofs from the literature: Studies have shown that exposure of hippocampal neurons to amyloid beta results in *inhibition* of canonical Wnt signaling that is important for cell proliferation, differentiation, adhesion, survival, and apoptosis. It is also shown that in hippocampal neurons amyloid beta cause induction of Dkk1 protein that acts as Wnt antagonist and *prevents* normal cell proliferation, differentiation, adhesion, survival, or apoptosis. In AD protein Dkk1 is elevated. Protein Dkk3 is also elevated and is present in plasma and cerebral fluid in AD. Neuronal impairment in brain and amyloid plaques prevents Wnt signaling and leads to induction of Dkk1 proteins that cause genes mutations or variations. Further, Apo lipoprotein E is found to be *risk factor* for AD that inhibits canonical Wnt signaling. Also, low density of receptor related LRP6 leads to genetic variations, and AD progression. Dkk1 protein also *reduce* amount of synaptic proteins inducing synaptic disassembly at pre and post synaptic sites. Clustering is found to be a *susceptibility factor* for late onset AD, and it *regulates* amyloid beta toxicity via Dkk1 driven induction of the non-canonicalWnt/PCP-JNK pathway, which *contributes* to tau phosphorylation and cognitive impairments. *Early event* in AD is synaptic failure which is caused by amyloid beta oligomers that are *soluble* to nerve cells and are responsible for synaptic pathology prior plaque deposition and neuronal death. Studies summarized in (2) support the idea that alterations in Wnt signaling pathway, are involved in modulation of synaptic development as well as in progression of AD. The activation of signaling pathways that crosstalk with Wnt pathway, support the neuroprotective potential of Wnt signaling cascade in AD.

In basal forebrain there is a loss of cholinergic neurons and due to cortical deficiencies in cholinergic neurotransmission there is impairment of cognitive functions and behavioral disturbances in AD. This is found after postmortem AD autopsy that disturbances are found in the metabolism of acetylcholine. Decrease of cholinergic neurons leads to the alteration of several proteins in the cholinergic system, such as decreased activity of acetylcholinesterase AChE and cholineacetyl transferase. Macromolecule found in synapsis interacts with amyloid beta and form complex which alters synaptic function in hippocampal neurons. Amyloid beta AChE complexes are more neurotoxic then amyloid beta alone depending on level of AChE, suggesting that AChE plays key role in neurodegenerative changes in AD, [2].

There are many proofs and data in the literature, and our aim here was to give clues how brain damage or impairment is caused by neurotoxicity of amyloid plaques and neurofibrillary tangles and more complex oligomers, and to show that above mentioned signaling pathways have protective or neuro protective effects in fighting these physical abnormalities, with inferring that possible cross talk with other mechanisms and pathways might influence signaling in pathway, transcription of genes, protein synthetized, cell functions or fate, and diseases that include Alzheimer disease as well. Impairment or damage of neural cells in brain cause functional and cognitive decline and in general loss of capabilities.

It is therefore important to investigate these pathways, and in this paper particularly Wnt signaling pathway. We focus on canonical Wnt/beta catenin signaling cascade, in order to show simple model, to set computational framework for analyzing more complex pathways and cross talks in further research.

Biochemical reactions in cytoplasm,

$$\begin{gathered}
D_i \Leftrightarrow D_a, \mathrm{k}_1, \mathrm{k}_2, \\
D_a + \mathrm{Y}_i \Leftrightarrow \mathrm{IC}_{\mathrm{xd}} \rightarrow D_a + \mathrm{Y}_a, \mathrm{k}_3, \mathrm{k}_4, \mathrm{k}_5, \\
Y_a + \mathrm{P} \Leftrightarrow \mathrm{IC}_{\mathrm{yp}} \rightarrow Y_i + \mathrm{P}, \mathrm{k}_6, \mathrm{k}_7, \mathrm{k}_8 \\
Y_i + \mathrm{x} \Leftrightarrow \mathrm{IC}_{\mathrm{xy}} \rightarrow Y_i + 0, \mathrm{k}_9, \mathrm{k}_{10}, \mathrm{k}_{11} \\
0 \rightarrow \mathrm{x}, \mathrm{k}_{12} \\
x \rightarrow 0, \mathrm{k}_{13}
\end{gathered}$$

Biochemical reactions in nucleus,

$$\begin{gathered}
D_{\mathrm{an}} + \mathrm{Y}_{\mathrm{in}} \Leftrightarrow X_{\mathrm{dyn}} \rightarrow D_{\mathrm{an}} + \mathrm{Y}_{\mathrm{an}}, \mathrm{k}_{14}, \mathrm{k}_{15}, \mathrm{k}_{16} \\
Y_{\mathrm{an}} + \mathrm{P}_n \Leftrightarrow \mathrm{IC}_{\mathrm{ypn}} \rightarrow Y_{\mathrm{in}} + \mathrm{P}_n, \mathrm{k}_{17}, \mathrm{k}_{18}, \mathrm{k}_{19} \\
Y_{\mathrm{in}} + \mathrm{x}_n \Leftrightarrow \mathrm{IC}_{\mathrm{xyn}} \rightarrow Y_{\mathrm{in}} + 0, \mathrm{k}_{20}, \mathrm{k}_{21}, \mathrm{k}_{22} \\
x_n \rightarrow 0, \mathrm{k}_{23} \\
x_n + \mathrm{TF} \Leftrightarrow \mathrm{IC}_x\mathrm{TF}_n, \mathrm{k}_{24}, \mathrm{k}_{25}
\end{gathered}$$

Biochemical reactions in Cytoplasm to Nucleus,

$$\begin{gathered}
D_a \Leftrightarrow D_{\mathrm{an}}, \mathrm{k}_{26}, \mathrm{k}_{27} \\
Y_a \Leftrightarrow Y_{\mathrm{an}}, \mathrm{k}_{28}, \mathrm{k}_{29}
\end{gathered}$$

6 Mathematical Model of Wnt Signalling Pathway

Next, we give mathematical model of Wnt/β catenin signalling pathway, [6]:

$$\dot{x}_1 = -k_1x_1 + k_2x_2 \tag{1}$$

$$\dot{x}_2 = k_1x_1 - (k_2 + k_{26})x_2 + k_{27}x_3 - k_3x_2x_4 + (k_4 + k_5)x_{14} \tag{2}$$

$$\dot{x}_3 = k_{26}x_2 - k_{27}x_3 - k_{14}x_3x_6 + (k_{15} + k_{16})x_{15} \tag{3}$$

$$\dot{x}_4 = -k_3x_2x_4 - k_9x_4x_{10} + k_4x_{14} + k_8x_{16} + (k_{10} + k_{11})x_{18} \tag{4}$$

$$\dot{x}_5 = -k_{28}x_5 + k_{29}x_7 - k_6x_5x_8 + k_5x_{14} + k_7x_{16} \tag{5}$$

$$\dot{x}_6 = -k_{14}x_3x_6 - k_{20}x_6x_{11} + k_{15}x_{15} + k_{19}x_{17} + (k_{21} + k_{22})x_{19} \tag{6}$$

$$\dot{x}_7 = k_{28}x_5 - k_{29}x_7 - k_{17}x_7x_9 + k_{16}x_{15} + k_{18}x_{17} \tag{7}$$

$$\dot{x}_8 = -\dot{x}_{16} = -k_6x_5x_8 + (k_7 + k_8)x_{16} \tag{8}$$

$$\dot{x}_9 = -\dot{x}_{17} = -k_{17}x_7x_9 + (k_{18} + k_{19})x_{17} \tag{9}$$

$$\dot{x}_{10} = k_{12} - (k_{13} + k_{30})x_{10} - k_9x_4x_{10} + k_{31}x_{11} + k_{10}x_{18} \tag{10}$$

$$\dot{x}_{11} = -k_{23}x_{11} + k_{30}x_{10} - k_{31}x_{11} - k_{20}x_6x_{11} - k_{24}x_{11}x_{12} + k_{25}x_{13} + k_{21}x_{19} \tag{11}$$

$$\dot{x}_{12} = -\dot{x}_{13} = -k_{24}x_{11}x_{12} + k_{25}x_{13} \tag{12}$$

$$\dot{x}_{14} = k_3x_2x_4 - (k_4 + k_5)x_{14} \tag{13}$$

$$\dot{x}_{15} = k_{14}x_3x_6 - (k_{15} + k_{16})x_{15} \tag{14}$$

$$\dot{x}_{18} = k_9x_4x_{10} - (k_{10} + k_{11})x_{18} \tag{15}$$

$$\dot{x}_{19} = k_{20}x_6x_{11} - (k_{21} + k_{22})x_{19} \tag{16}$$

Conservation laws are,

$$0 = (x_1 + x_2 + x_3 + x_{14} + x_{15}) - c_1 \tag{17}$$

$$0 = (x_4 + x_5 + x_6 + x_7 + x_{14} + x_{15} + x_{16} + x_{17} + x_{18} + x_{19}) - c_2 \tag{18}$$

$$0 = (x_8 + x_{16}) - c_3 \tag{19}$$

$$0 = (x_9 + x_{17}) - c_4 \tag{20}$$

$$0 = (x_{12} + x_{13}) - c_5 \tag{21}$$

Some research on biomolecules and signalling pathways in Alzheimer disease is analysed in [7–9]. Further, NfkB and MAPK signalling pathways that are also important for biological functions are modelled in [10, 11].

Interesting biological phenomena in molecular mechanisms, signalling pathways or biological networks can also be approximated or modelled as singular or impulsive phenomena. Therefore system theory developed for singular and impulsive systems [12–14] can be used. Results are given in [15]. For example, we could state general dynamical model by including conservation laws, to obtain differential – algebraic model, i.e. singular system. Thresholds in biological models can be modelled as impulsive phenomena.

7 Simulation Results

In order to obtain simulation results we give table of initial conditions and rates of reactions that we will use in simulations (Table 2).

Table 2. Rate of reactions

Rate of reaction	Value	Rate of reaction	Value
k_1	92.331732	k_{17}	0.61699064
k_2	0.86466471	k_{18}	0.61699064
k_3	79.9512906	k_{19}	37.913879
k_4	97.932525	k_{20}	0.86466471
k_5	1	k_{21}	0.86466471
k_6	3.4134082	k_{22}	0.99326205
k_7	0.61409879	k_{23}	0.99326205
k_8	0.61409879	k_{24}	1
k_9	3.4134082	k_{25}	5.9744464
k_{10}	0.98168436	k_{26}	1.7182818
k_{11}	0.98168436	k_{27}	1.7182818
k_{12}	4.7267833	k_{28}	1.7182818
k_{13}	0.17182818	k_{29}	1.7182818
k_{14}	0.68292191	k_{30}	0.55950727
k_{15}	1	k_{31}	1.0117639
k_{16}	3.2654672		

Conserved quantities are given in Table 3.

Table 3. Conserved quantities

Conserved quantities	Value
c_1	4.9951026
c_2	16.4733784
c_3	1.6006340
c_4	1.2089126
c_5	2.7756596

Next, we state set of initial conditions.

Table 4. Initial conditions

Variable	Initial value	Variable	Initial value
x_1	4.9951	x_{11}	0
x_2	0	x_{12}	2.77566;
x_3	0	x_{13}	0
x_4	16.4734	x_{14}	0
x_5	0	x_{15}	0
x_6	0	x_{16}	0
x_7	0	x_{17}	0
x_8	1.60063	x_{18}	0
x_9	1.20891	x_{19}	0
x_{10}	0		

Initial conditions are chosen in such a way that certain quantities are present for inactive Disheveled, inactive destruction complex, phosphatase in cytoplasm, phosphatase in nucleus, and transcription factor in nucleus. Other quantities and intermediate complexes are at initial time equal to zero (Fig. 1). Therefore, quantities chosen for initial conditions given in Table 4 straightforwardly correspond to conservation quantities given in Table 3.

From the simulation results above we see that Disheveled inactive in cytoplasm and active in cytoplasm and nucleus rise from initial value to steady state. Rise of inactive Disheveled is due to inactivation from active Disheveled that is produced when Wnt signaling is on. Destruction complex inactive in cytoplasm decrease from initial state to zero, since it reacts with active Disheveled to form intermediate complex and produce active destruction complex. It is consumed as more active Disheveled is formed. Active destruction complex both in cytoplasm and nucleus and inactive destruction complex in nucleus rise from zero to steady state, since active Disheveled is produced that react with destruction complex. Phosphates both in cytoplasm and nucleus decrease from initial state to steady state, since they are consumed in biological process with active destruction complexes to form intermediate complexes, and inactivate destruction complexes. Beta catenin has impulse rise in cytoplasm to steady state, due to production, and in nucleus also rise from zero to steady state. It also degrades but rate of

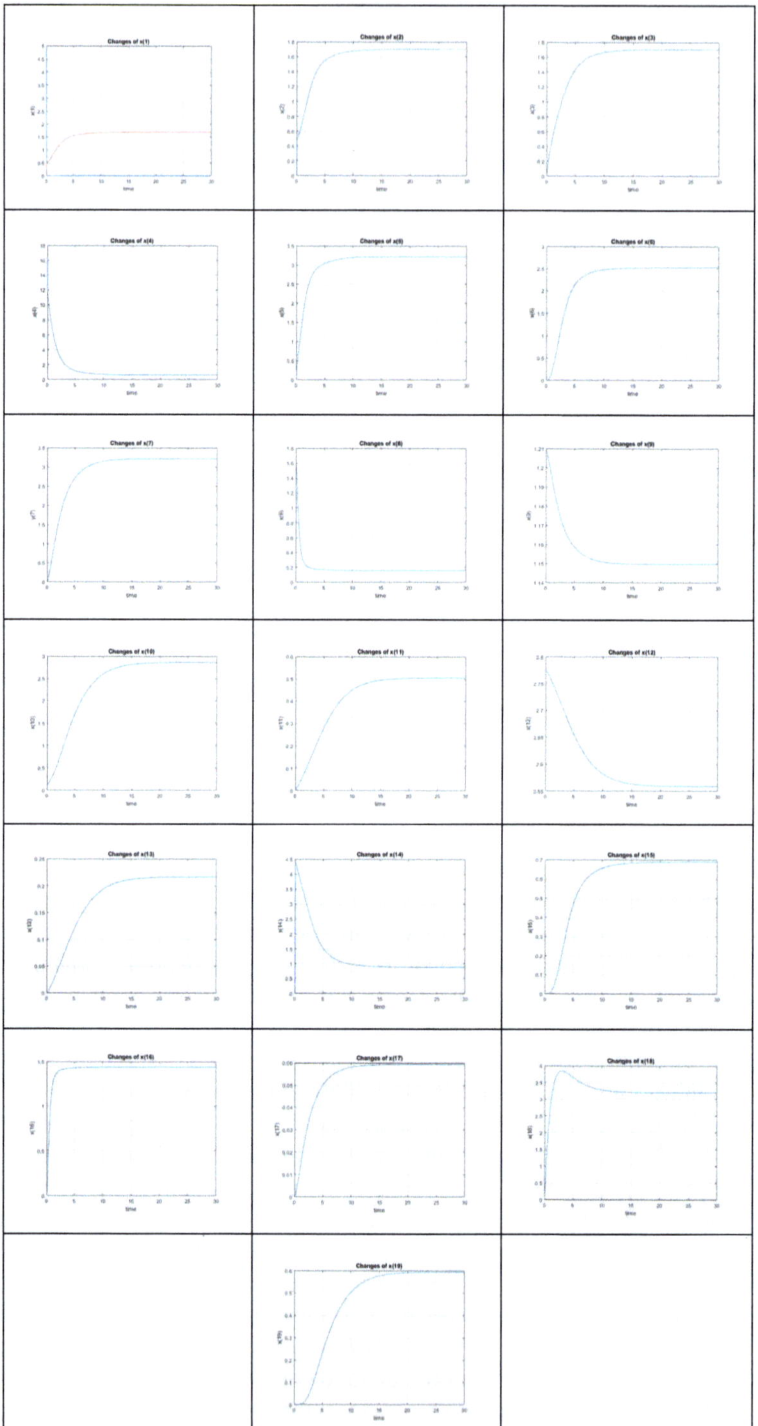

Fig. 1. Simulation results

degradation k_{13} is lower than rate of production k_{12}. Transcription factor decrease from initial state to steady state, since it is consumed for transcription due to release of beta catenin that promotes process of transcription. Intermediate complex of beta catenin and disheveled in cytoplasm decrease, while complexes of beta catenin-transcription factor, beta catenin disheveled in nucleus, complex of destruction complex and phosphate in cytoplasm and nucleus rise. Intermediate complex of beta catenin and destruction complex rise, have transient behavior and reach steady state. Intermediate complex of beta catenin and destruction complex in nucleus rise.

8 Analysis of Mathematical Model

Mathematical model we have obtained is nonlinear with 19 states and 31 rate reactions. Since it is nonlinear it can show multiple responses and multistability of Wnt signalling pathway under certain different conditions. By investigating the model in Mathematica software we have obtained two sets of equilibrium states that can lead to different cell functions (for example, stem cells proliferation or differentiation) depending of the chosen equilibria of the model. Equilibrium points are functions of rate of reactions as parameters, what leads to different steady states obtained. Further, it can be performed bifurcation analysis of biochemical rates parameters in order to show changes in system dynamics that leads to different responses, like bistability, switch-like transition, hysteresis, or graded response [2]. Model resembles two steady states switching between committed state and stem like state. In the bi stable regime the low level of gene transcription is associated with a committed cell state and the high level with a stem-like phenotype over long time periods. As the value of a parameter, for example β-catenin shuttling into the cytoplasm k_{25}, decreases below a threshold, the bifurcation diagram predicts that cells will differentiate. If the shuttling rate was adequately increased these cells would dedifferentiate to a stem-like state. If the parameter regime were known, bifurcation analysis and singularity analysis could also predict parameters governing reversible and irreversible behaviour, like parameter k_5 that leads to irreversible behaviour. This is a topic of further research.

Nonlinear systems in general exhibit multiple equilibrium points, limit cycles - oscillations of constant amplitude and frequency, chaos phenomena: randomness, complicated steady state behaviours, and multiple modes of behaviour. For the model presented we have found two sets of equilibrium states, and by performing linearization around equilibrium point, we have obtained linear matrix that we have tested for eigenvalues in order to determine stability, but we have found three positive equilibria states with other ones being generally negative or zero eigenvalues. From First or indirect method of Lyapunov we could not conclude stability of matrix of linearized system and therefore nonlinear system, although simulation results shows that states attained steady values.

In mathematical models governing biochemical processes or signalling pathways there are present large number of system states and is therefore hard to perform analytical study, rather we use computational approach, numerical, stability, and bifurcation studies in order to show graphically how system states are changed due time, dependence of variables, or parameter values that lead to different state responses, or state changes depending of parameters. Even numerical study becomes impossible for large number of system states, since time needed for simulation is large, and hardware requirements for components are more demanding. This is a problem that follows biochemical modelling, simulation, and dynamical analysis of signalling pathways.

Further research will focus on more detailed model of signalling pathway including other key actors. We will consider cross talk with other pathways, and analyse each of them. We will attempt to show dependence of signalling pathway actors with amyloid beta protein and plaques and tau proteins neurofibrily tangles. We will perform extensive research of the literature in order to connect and find clues from the results already proved, and try to explain them in line with mutations, variations of genes, dysfunctions of pathways, irregularities, abnormal proteins, etc. In order to decipher connections with Alzheimer disease pathology.

9 Conclusion

In this paper we have explained what the Alzheimer disease is, what are signs and pathological abnormalities that follow this disease, main symptoms, causes that can be environmental, biological, or genetics, or interfered with specific other diseases that can trigger AD earlier. We have explained possible crosstalk of different signalling pathways that are found to be key acters in Alzheimer disease: AMPK, mTOR, Sirtu1, PCB-1, Wnt. We left complex analyse for further research and we focused on canonical Wnt/beta catenin signalling pathway, since we are interested to find connections and clues between neural cell damage or impairment in brain, accumulation of amyloid plaques and neurofibrillary tangles, amyloid beta protein and their complexes, tau protein and explain taupatologies that appear in Alzheimer disease, and their connection with signalling pathways and other compounds and molecular mechanisms and their crosstalk. We are particularly interested to set computational model consisted of set of biochemical reactions that are based on molecular mechanism beneath, and mathematical model, for which we are interested in order to find responses of states or components in signalling pathway due time, and their responses. Of our interest is beta catenin and beta catenin destruction intermediate complex that triggers gene transcription and determine cell function between two possible states: stem-cell or differentiation. Thorough analytical and numerical investigation is topic of further research, and this paper presents initial base with concept, framework and results that will be used in future work.

References

1. Alzheimer disease. http://www.memorylossonline.com/glossary/alzheimer.html
2. Godoy, J.A., Rios, J.A., Zolezzi, J.M., Braidy, N., Intestrosa, N.C.: Signalling pathways crosstalk in Alzheimer's disease. Cell Commun. Signal. **12**, 23 (2014)
3. Vallee, A., Lecarpentier, Y.: Alzheimer disease: crosstalk between canonical Wnt/beta catenin pathway and PPAR alpha and gamma. Front. Neurosci. **10** (2016). Article no. 459. https://doi.org/10.3389/fnins.2016.00459

4. Cadigan, K.M.: Wnt-Beta-catenin signaling. Magazine R943
5. MacLean, A., Rosen, Z., Byrne, H.M., Harrington, H.A.: Parameter free methods distinguishWnt pathway models and guide design of experiments. PNAS **119**(9), 2652–2657 (2015)
6. Gross, E., Harrington, H.A., Rosen, Z., Sturmfeels, B.: Algebraic systems biology: a case study for the Wnt pathway. arXiv:1502.03188v1 (2015)
7. Mitrasinovic, O., Kablar, N.A.: Computational approaches in preclinical diagnostics and prognosis for Alzheimer disease. Alzheimer Dementia **13**(7), 1005–1006 (2017)
8. Mitrasinovic, O., Kablar, N.A.: Emerging computational strategies identify MyD88 as downstream target in interleukin-1α induced signal transduction in Alzheimer's disease. Alzheimer Dementia **5**(4), 21–22 (2009)
9. Mitrasinovic, O., Kablar, N.A.: P3-366: indirect neuroprotective effects of interleukin-1α in the hippocampal ex vivo organotypic co-culture model. Alzheimer Dementia **4**(4), 628–629 (2008)
10. Kablar, N.A.: Mathematical model of IL -1- NfkB biological module. Glob. J. Math. Sci. **1** (1) (2012)
11. Kablar, N.A.: MAPK module: biological basis, structure, mathematical model and dynamical analyse. In: Proceedings of the 19th International Symposium on Mathematical Theory of Networks and Systems – MTNS 2010, Budapest, Hungary (2010)
12. Haddad, W.M., Chellaboina, V., Kablar, N.A.: Nonlinear impulsive dynamical systems, part I: stability and dissipativity. Int. J. Control **74**, 1631–1658 (2001)
13. Haddad, W.M., Chellaboina, V., Kablar, N.A.: Nonlinear impulsive dynamical systems, part II: feedback interconnections and optimality. Int. J. Control **74**, 1659–1677 (2001)
14. Haddad, W.M., Kablar, N.A., Chellaboina, V.: Optimal disturbance rejection control for nonlinear impulsive dynamical systems. Nonlinear Anal. Theory Methods Appl. **62**(8), 1466–1489 (2005)
15. Kablar, N.A., Debeljković, D.: Singularly Impulsive Dynamical Systems and Applications in Biology. Scientific Monograph (2015). ISSBN 978-86-7083-849-9

Identification and Recognition of Vehicle Environment Using Artificial Neural Networks

Darko Jocic(✉), Velimir Cirovic, and Dragan Aleksendric

Faculty of the Mechanical Engineering, University of Belgrade, 11120 Belgrade, Serbia
djocic@mas.bg.ac.rs

Abstract. Object detection using deep learning over the years became one of the most popular methods for implementation in autonomous systems. Autonomous vehicle requires very reliable and accurate identification and recognition of surrounding objects in real traffic environments to achieve decent detection results. In this paper, special type of Artificial Neural Network (ANN) named Convolutional Neural Network (CNN) was used for identification and recognition of surrounding objects in real traffic. The new model based on CNN was trained and developed to be able to identify and recognize 4 different classes of objects: cars, traffic lights, persons and bicycles. The developed model has shown 94.6% accuracy of object identification and recognizing on the test set.

Keywords: Artificial Neural Networks · Convolutional Neural Network
Object detection · Vehicles environment

1 Introduction

For decades, object recognition and detection have been important problems in real-life applications of autonomous vehicles [1]. Vehicle environment is very complex and dynamically changeable. Every part of the street, especially crossroads, can be often full of complex and dynamic objects such as pedestrians, traffic lights, different types of vehicles, buildings, trees, etc. In general, besides objects shape and complexity, the illumination, light condition, weather condition, partial occlusion, presence of shadow, etc. are one of the major problems in object detection field. Making preconditions for an autonomous vehicle to run safely through such complex environment is a very demanding task. Bearing in mind that the active safety and comfortability of modern vehicles has become a major concern in today's research activities, recently developed systems for the intelligent braking control [2–4], for example, must be involved and integrated with the system for identification and recognition of the vehicles environment. It is important for future autonomous vehicles (AV) in order to avoid obstacles and prevent possible crashes.

According to [5–7], ANNs as one of the artificial intelligence technique offer possibilities to develop such system able to identify and recognize objects in real traffic environments. There are different approaches to the identification and object recognition problem. For example, two types of detection systems have been shown in [8], for road detection and on-road detection. The on-road detection mainly concerns

N. Mitrovic et al. (Eds.): CNNTech 2018, LNNS 54, pp. 208–219, 2019.
https://doi.org/10.1007/978-3-319-99620-2_16

vehicles and pedestrians class object, due to various types, shapes and size of objects. These two methods are not robust and not general enough for the application of autonomous vehicles [9, 10]. The research presented in [11] is based on the CNN method, and it is shown that presence of illumination, light condition and occlusion highly affect detection results. Deep Neural Networks (DNN) have been proposed in [12, 13] to classify vehicles. Moving vehicle detection algorithm based on CNN was presented in [14] using the vehicles' frame difference to detect different types of vehicles.

In this paper, the deep learning method has been employed in order to develop a dynamic model for object detection and recognition during vehicle driving in real traffic environment. This model is based on special type of ANNs named Convolutional neural networks (CNNs). Due to simplicity and performance that CNN offers, the dynamic model has been developed using wide spectrum of different class of objects such as cars, traffic lights, persons and bicycles. The developed neural model is able to detect and identify obstacles in dynamically changing vehicle environment conditions. Here proposed approach in synergy with other important tasks (intelligent vehicle braking, for instance) represents a starting point and crucial part towards autonomous vehicle driving.

2 Object Detection Using Convolutional Neural Networks

In general, ANNs are based on the neurobiological basis of the human brain. Biological neurons are slower than modern electric circuits, however, the brain is faster than any computer due to large number of neurons that work parallel. ANNs can learn, adapt to environment of the problem being considered and can establish a "model" in situations where rules are unknown and/or are vague and/or incomplete [15]. CNNs are special type of ANNs inspired by the biological visual cortex of humans. Visual cortex is responsible for processing visual information in the brain. It has sequence of areas that process the information in a low to high abstraction level. The study of visual cortex shows that the neurons presented in it get activated by stimulations generated by localized fields. Linear filtering in image processing is performed through Convolutional in spatial domain (or element-wise multiplication in frequency domain). Idea behind CNN is to learn these filters in a data-driven manner [16]. CNNs are variants of multilayer perceptron. CNNs consist of four types of layers: Convolutional layers, Pooling (Subsampling) layers, Non-linear layers and Fully Connected layers. Images, which can be considered to be stationary, have almost the same 67 statistics in the different smaller parts of the image. This means that if we learn feature extractors for one part of the image, the same feature extractors can be used to extract features from other parts of the image. This idea is the main motivation behind the Convolutional layer, as the main part of CNN. However, to reduce the computational complexity and to avoid over-fitting there is a need to summarize the features, which leads to the idea of having subsampling [16].

The model used in this paper is based on the "AlexNet" pre-trained model [17]. This model has been adopted for object recognition and identification task. It is retrained with additional set of images to provide desired. Such model includes several pre-trained networks. Most of these have been trained on the ImageNet dataset, which has 1000 object categories and 1.2 million training images, according to [17]. The "AlexNet" is based on CNN trained on two different GPUs, containing 23 layers, five Convolutional layers, three max-pooling layers, and three fully connected layers. The architecture of this CNN model is shown in Fig. 1.

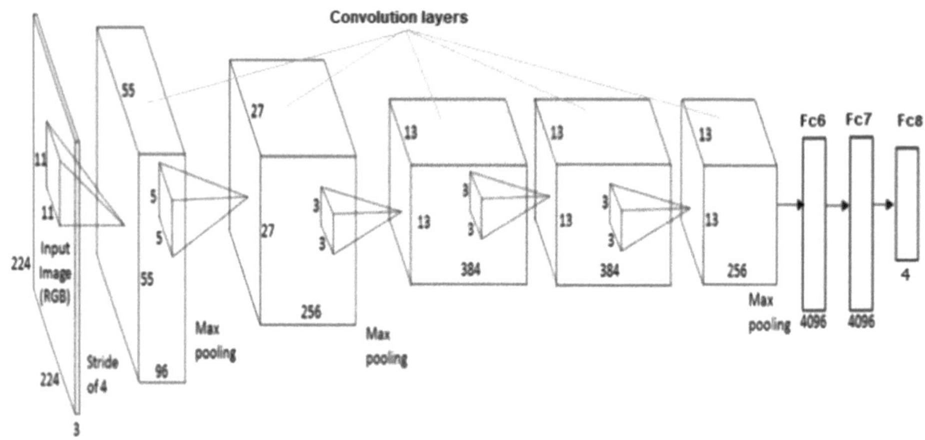

Fig. 1. The architecture of CNN model having 5 convolutional and 3 max pooling (sub-sampling) layers [18].

2.1 Convolutional Layers

The Convolutional layer of CNN model uses multiple learned filters to obtain multiple filter maps detecting low-level filters. Convolutional filter kernel weights are decided through the training process. The first Convolutional layer extracts low-level features like edges, lines and corners. Higher-level Convolutional layers extract higher-level features, as it was shown in [19]. Acquired pre-processed segmented image is resized to $227 \times 227 \times 3$ that represent width, length and the three-color channels in RGB colour space. Then 96 kernels are convolved with the input pre-processed image for extraction low level features and output 96 sub images of size 55×55. Figure 2 shows an example of filters learned by Krizhevsky, as explained in [17]. Each of the 96 filters shown in Fig. 2 has size of $[11 \times 11 \times 3]$, and each one is shared by the $55 * 55$ neurons in one depth slice. These kernels attempt to extract the edges. These 96 kernels divided into two sets, each of them 48 kernels run on specific Graphical Processing Unit (GPU) [18]. The first set is colour-agnostic and second set is colour-specific.

Figure 3 illustrates the process of 3D convolutional, used in CNN model. The input image has size $227 \times 227 \times 3$ and is convolved with 96 kernels, each of size $11 \times 11 \times 3$ separately. Convolutional with one kernel produces one output feature, and with 96 kernels independently produces 96 features.

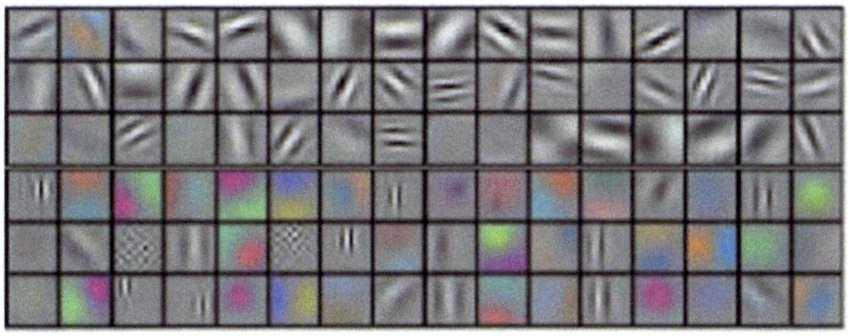

Fig. 2. Convolutional kernel size 11 × 11 × 3 learned by the first convolutional layer of CNN model on the 227 × 227 × 3 input image [17].

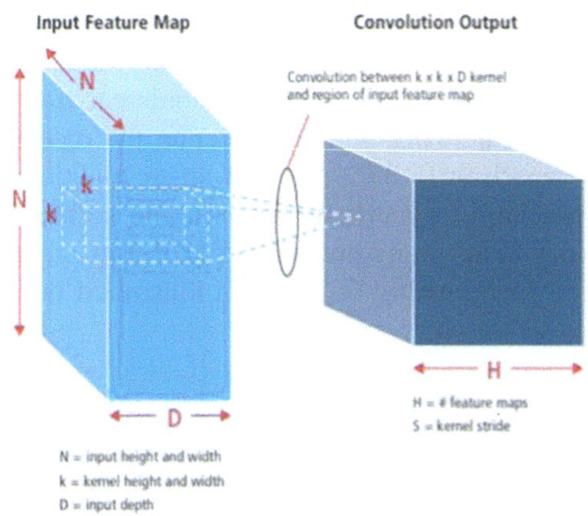

Fig. 3. Pattern of 3D convolutional [20].

2.2 Pooling (Subsampling) Layers

The pooling (subsampling) layer reduces the resolution of the features. The main task of this layer is to make feature representations smaller and more manageable and it makes the features robust against noise and distortion [20]. Figure 4 shows how the pooling layer down-samples the volume spatially, showing the effect of pooling one of the 96 convolved images generated after first convolutional layer. The input volume of size [55 × 55 × 96] is pooled with filter size of 2 and stride of 2 into an output volume of size [27 × 27 × 96], while the depth is preserved. Figure 4 also contains an example how max pooling operation works. The input size of 4 × 4 is divided into four non-overlapping matrices of size 2 × 2, while the output value is the maximum value out of four values in the 2 × 2 matrixes.

2.3 Non-linear Layers

CNNs in particular rely on a non-linear "trigger" function to signal distinct identification of likely features on each hidden layer. CNNs may use a variety of specific functions such as rectified linear units (ReLUs) and continuous trigger (non-linear)

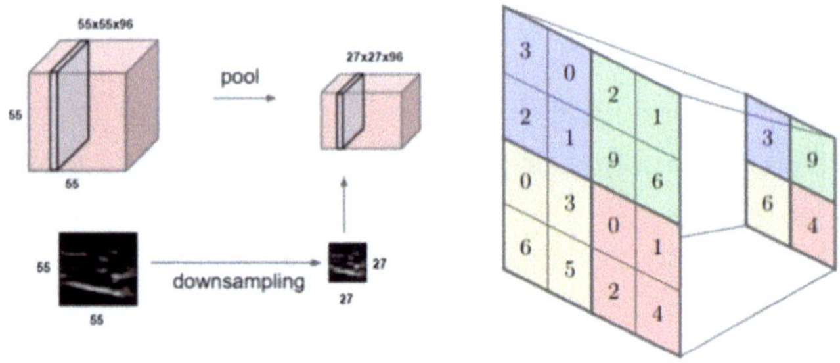

Fig. 4. Effect of pooling (subsampling) layer of convolved image (left), and example of max pooling (right), [21].

functions. A ReLU implements the function $y = max\ (x,0)$, so the input and output sizes of this layer are the same. It increases the nonlinear properties of the decision function and of the overall network without affecting the receptive fields of the convolutional layer. In comparison to the other non-linear functions used in CNNs (e.g., hyperbolic tangent, absolute of hyperbolic tangent, and sigmoid), the advantage of a ReLU is that the network trains many times faster [20]. ReLU "trigger" functions were used in this model. The functionality of ReLU is illustrated in Fig. 5 with its transfer function plotted above the arrow.

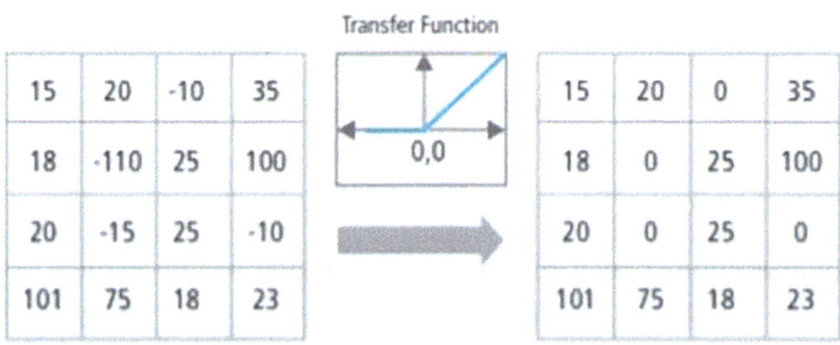

Fig. 5. Representation of ReLU functionality according to [20].

2.4 Fully Connected Layers

Fully connected layers are often used as the final layers of a CNN model. These layers mathematically sum a weighting of the previous layer of features, indicating the precise mix of "ingredients" to determine a specific target output result. In case of a fully

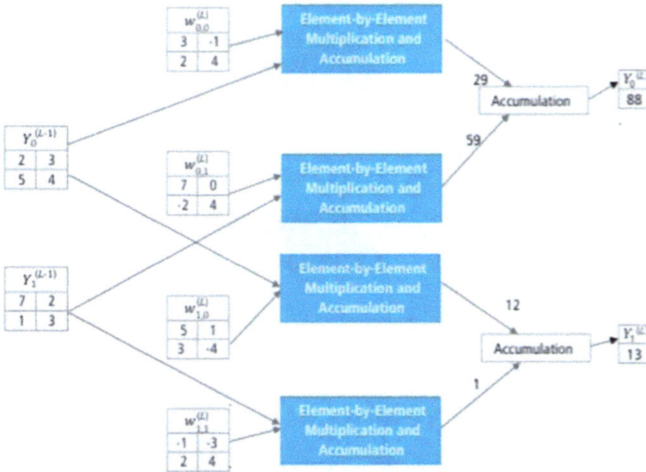

Fig. 6. Processing of a fully connected layer [20].

connected layer, all the elements of all the features of the previous layer get used in the calculation of each element of each output feature [20]. Figure 6 illustrates how the features in the fully connected layer are calculated.

Selecting which of deeper layers to choose for feature extraction is a design specific. There are three fully connected layers (fc6, fc7 and fc8) that combine features from all previous layers. Last layer (fc8) represent classification layer (4 classes). In the fc7 features are extracted. These features are calculating out of 4096 neurons, combining the lower level features with connected weights.

3 Classifying Objects Using the Faster Convolutional Neural Network (Faster R-CNN)

Object detection system used in this paper, called Faster R-CNN, is composed of two modules. The first module is a deep fully convolutional network that proposes regions and the second module is the Fast R-CNN detector that uses the proposed regions. The entire system is a single, unified network for object detection (see Fig. 7). Faster R-CNN is a single, unified network for object detection. The Region Proposal Network (RPN) module serves as the 'attention' of this unified network, or using the recently popular terminology of neural networks with 'attention' mechanisms, the RPN module tells the Fast R-CNN module where to look [22].

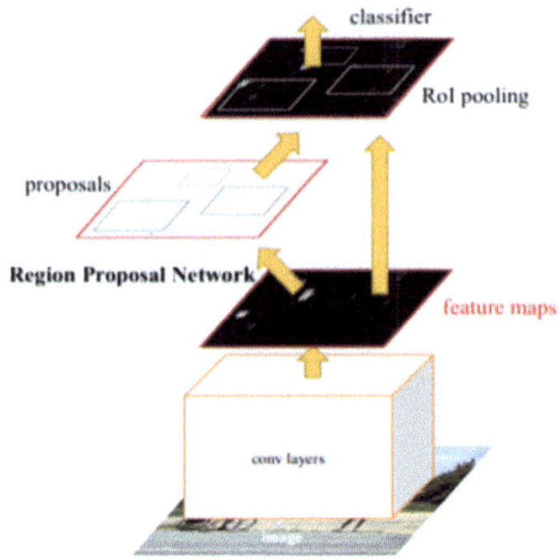

Fig. 7. Object detection system based on the Faster R-CNN [22].

4 Results and Discussion

4.1 Dataset Description

Images used to train CNN model were coming from different sources such as ImageNet and Coco images [23, 24]. Around 8000 different types of images were collected and equally divided into 4 categories namely as Cars, Persons, Bicycles and Traffic Lights. Figure 8 shows the example of image used to train Faster R-CNN classifier. For training Faster R-CNN classifier 400 images have been used, equally divided in 4 categories. Dataset used for training CNN model contains 70% of total amount of images equally separated into categories mentioned above, while the rest percentage of images has been used as test set to validate prediction performance of CNN model developed in this paper. Images used to test Faster R-CNN detector (classifier) have been used as the new set of images, which does not belong to the set that has been used to train or test the developed CNN model.

4.2 Test Results

Tests have been carried out to validate the efficiency of the proposed neural model. In Figs. 9, 10, 11 and 12 are shown results of Faster R-CNN detector in various traffic situations. CNN model achieved accuracy of 94.6% on test set. Images used for evaluating detector are from [23, 24].

In Fig. 9 are shown detection results of 2 out 3 vehicles on parking site. Traffic lights in background are not recognized, which clearly confirms that Faster R-CNN detector require more data in order to provide better detection. From the sight of the CNN model, recognized objects were identified correctly. On Figs. 10 and 12 there is similar problem as on Fig. 9 that not all the objects were detected by detector, while on Fig. 11 decent detection results were achieved, detector/CNN model wise.

Fig. 8. Example of images used to train Faster R-CNN classifier [24].

Fig. 9. Results of Faster R-CNN detector, recognized and identified Cars on parking site.

Fig. 10. Identified and recognized Pedestrians, Traffic Light and Bicycle.

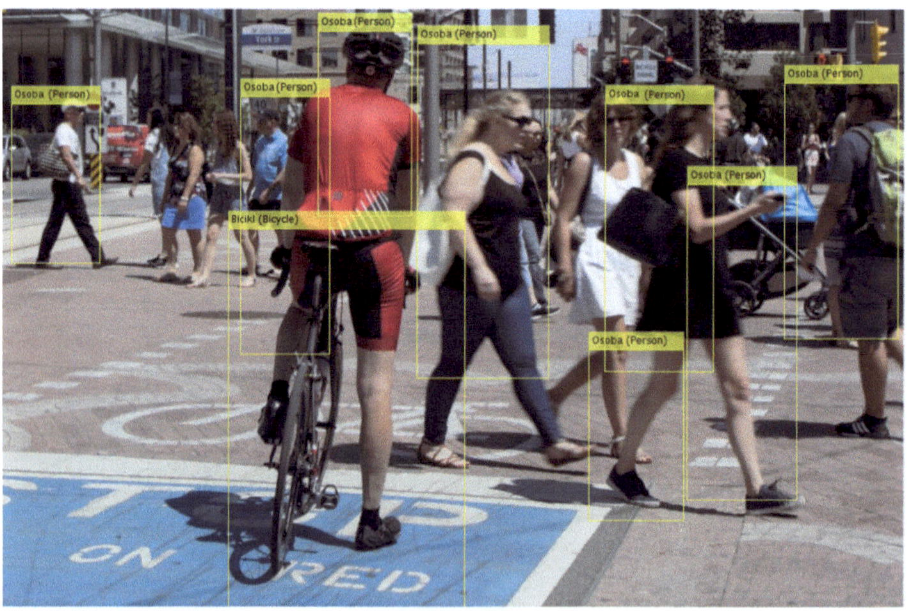

Fig. 11. Identified and recognized Pedestrians and Bicycle

Fig. 12. Identified and recognized Pedestrians, Car and Traffic Light.

5 Conclusion

In this paper, a combined system of CNN and Faster R-CNN has been used, able to accomplish object and/or obstacle recognition and detection task in dynamic environment. In real environment, the appearance of objects varies due to large number of influencing factors, like variations of light condition, illumination, weather condition, partial occlusion, presence of shadow and surrounding background clutters. All mentioned appearances strongly influence the performance of developed CNN model, making the recognition and detection process more difficult. Here developed model for object detection in dynamic environment, using special type of ANN (CNN), has been trained on set of 8000 images equally divided into four categories, while Faster R-CNN detector is trained with set of 400 images, also equally divided into four categories. Results achieved on test set with these two models are 94.6% of detected and recognized objects based on relatively small amount of data used to train the detector (Faster R-CNN). However, achieved results are promising in regard to amount of data being used, but there is still more space to develop this model further to be capable to identify and recognize more types of objects.

References

1. Aysxegul, U., Yakup, D., Cuneyt, G.: Object recognition and detection with deep learning for autonomous driving applications. Simulation **93**, 759–760 (2017). https://doi.org/10.1177/0037549717709932
2. Aleksendrić, D., Jakovljević, Ž., Ćirović, V.: Intelligent control of braking process. Expert Syst. Appl. **39**, 11758–11765 (2012). https://doi.org/10.1016/j.eswa.2012.04.076

3. Ćirović, V., Aleksendrić, D., Smiljanić, D.: Longitudal wheel slip control using dynamic neural networks. Mechatronics **23**, 135–146 (2013). https://doi.org/10.1016/j.mechatronics.2012.11.007
4. Ćirović, V., Aleksendrić, D.: Adaptive neuro-fuzzy wheel slip control. Expert Syst. Appl. **40**, 5197–5209 (2013). https://doi.org/10.1016/j.eswa.2013.03.012
5. Tohidul, I.K., Ram, G.R.: Real-time (vision-based) road sign recognition using an artificial neural network. Sensors **17**, 853 (2017). https://doi.org/10.3390/s17040853
6. Lorsakul, A., Suthakorn, J.: Traffic sign recognition using neural network on OpenCV: toward intelligent vehicle/driver assistance system. In: The 4th International Conference on Ubiquitous Robots and Ambient Intelligence (URAI 2007) (2007)
7. Lazrus, A., Choubey, S.: A robust method of license plate recognition using ANN. IJCSIT **2**, 1494–1497 (2011)
8. Pendleton, S.D., Andersen, H., Du, X., Shen, X., Meghjani, M.: Perception, planning, control, and coordination for autonomous vehicles. Machines **5**, 1–54 (2017). https://doi.org/10.3390/machines5010006
9. Zhao, Z., Zhou, L., Zhu, Q., Luo, Y.: A review of essential technologies for collision avoidance assistance systems. Adv. Mech. Eng. **9**, 1–15 (2017). https://doi.org/10.1177/1687814017725246
10. Mukhtar, A., Xia, L., Tang, T.B.: Vehicle detection techniques for collision avoidance systems: a review. IEEE Trans. Intell. Transp. Syst. **16**, 2318–2338 (2015). https://doi.org/10.1109/TITS.2015.2409109
11. Dheekonda, R.S., Panda, S., Khan, M.D.N., Hasan, M., Anwar, S.: Object detection from a vehicle using deep learning network and future integration with multi-sensor fusion algorithm. SAE Technical Paper No. 2017-01-0117 (2017). https://doi.org/10.4271/2017-01-0117
12. Zhou, Y., Cheung, N.: Vehicle classification using transferrable deep neural. In: Submitted for Presentation and Publication at the 2017 TRB Annual Meeting (2017)
13. Zhou, Y., Nejati, H., Do, T., Cheung, N., Cheah, L.: Image-based vehicle analysis using deep neural network: a systematic study. In: 2016 IEEE International Conference on Digital Signal Processing (DSP) (2016)
14. Ullah, I., Lee, H.: Moving vehicle detection and information extraction based on deep neural network. In: International Conference on Image Processing, Computer Vision, and Pattern Recognition—IPCV 2017 (2017)
15. Miljković, Z., Aleksendrić, D.: Veštačke neuronske mreže - Zbirka rešenih zadataka sa izvodima iz teorije. Univerzitet u Beogradu - Mašinski fakultet (2009). (in Serbian)
16. Prasoon, A.: Deep feature learning and cascaded classifier for large scale data. Ph.D. thesis, Department of Computer Science, Faculty of Science, University of Copenhagen (2014)
17. Krizhevsky, A., Sutskever, I., Hinton, GE.: ImageNet classification with deep convolutional neural networks. In: Proceeding NIPS 2012, Proceedings of the 25th International Conference on Neural Information Processing Systems (2012)
18. Ovtcharov, K., Ruwase, O., Kim, J., Fowers, J., Strauss, K., Chung, E.S.: Accelerating deep convolutional networks using specialized hardware. In: Microsoft Research (2015)
19. Srinivas, M., Roy, D., Mohan, C.K.: Discriminative feature extraction from X-ray images using deep convolutional neural networks. In: Proceedings of the 2016 IEEE International Conference on Acoustics, Speech and Signal Processing (ICASSP) (2016)
20. Hijazi, S., Kumar, R., Rowen, C.: Using Convolutional Neural Networks for Image Recognition. Cadence Design Systems Inc., California (2015)

21. CS23: Convolutional Neural Networks for Visual Recognition. http://cs231n.github.io/Convolutionalal-networks/
22. Ren, S., He, K., Girshick, R., Sun, J.: Faster R-CNN: Towards Real-Time Object. Computer Vision and Pattern Recognition (cs.CV), Ithaca (2016)
23. ImageNet. http://www.image-net.org/
24. COCO-Common Objects in Context. http://cocodataset.org/#home

Model Predictive Control of a Medical Robotic System

Ivan Buzurovic(✉)

Harvard Medical School, Harvard University, Boston, MA 02115, USA
ibuzurovic@bwh.harvard.edu

Abstract. One of the most challenging phases in interstitial brachytherapy is the placement of the needles. In these medical procedures, the needles are inserted inside the tissue to guide the positioning of the radioactive sources. The low dose-rate (LDR) radioactive sources are placed inside the tissue permanently, whereas a radioactive source in the high dose-rate (HDR) brachytherapy is temporarily placed in the desired positions so that the delivery of the prescription dose to the clinical targets can be achieved. Therefore, it is important to develop a robust and sophisticated tool that can perform the automatic needle placement with a high level of accuracy for different medical procedures and conditions. In this study, we propose a novel concept for the automatic needle insertion using a new miniature automated robotic system. The mathematical model of this system was derived, allowing the implementation of the model predictive control (MPC) that can be used to govern the mechanism. The purpose of this approach was to minimize the lateral components of the generalized reactive force which is responsible for the tissue displacement and, consequently, for the needle deflection.

Keywords: Robotic system · Model predictive control · Needle insertion

1 Introduction

In contemporary brachytherapy, the accurate positioning of the needles into the predefined locations is a challenging and complex task due to a variety of reasons [1]. The precision and reproducibility of the needle placement in manual brachytherapy procedures are highly dependent on the experience and dexterity of the physicians [2]. Some of the major problems that appear during the needle insertion in brachytherapy are caused by the displacement and deformation of the soft tissue. Prostate deformation as well as possible calcifications within the gland or denser tumour tissue can cause needle deflection and needle clustering, which is hard to track and estimate. The lack of spatial resolution in ultrasound imaging or lack of the real-time needle position estimation in interstitial brachytherapy makes the implementation of the automatic corrections especially difficult. Needle deflection and displacement can result in significant dosimetric discrepancies in the clinical practice if the position error was not assessed and encountered. For instance, if the needle placement accuracy was ± 1 mm, there

N. Mitrovic et al. (Eds.): CNNTech 2018, LNNS 54, pp. 220–230, 2019.
https://doi.org/10.1007/978-3-319-99620-2_17

exists a dose variation between 58% and 274% for a 40.7 cGy cm^2/h source of a high dose-rate (HDR) after loader at the point located 2 mm distal to the tip of the needle, as previously reported in [3].

Several researchers have analysed these problems. Thus, a variety of approaches were proposed to improve the insertion strategies [2, 4–11]. Some of the suggested methodologies included the evaluation of different trajectories (such as curvilinear or flexible insertions), correlation of the tissue deformation and the infinitesimal force per tissue displacement for the real-time updates of the needle trajectories, or different models capable of predicting the deformation of the tissue caused by needle insertions, etc. However, most these strategies have remained academic. Recently, several authors have reported the design, investigation and development of the image-guided robotic-based systems for the brachytherapy and biopsy procedures that included automatic needle insertions [3, 12–20]. The purpose of the listed systems was to accurately place the needles into the desired locations using robot assisted automatic or semi-automatic insertions. To the best of our knowledge, none of these robotic systems is in clinical use.

Considering the significance of the topic, we propose a novel approach to the automated needle insertion. Furthermore, we outlined the requirements for the development of a portable handheld (n + 2) degree-of-freedom (DOF) automatic device for needle insertion. This device can incorporate the proposed insertion strategies with the goal to insert the needle with a high level of accuracy. To achieve this goal, we report the detailed mathematical model for the proposed system. The main purpose of the mathematical model approach was twofold: (a) the mathematical model should accurately represent the relevant physical phenomena that significantly influence the system dynamics of such device, and (b) the mathematical model should be suitable for implementation in the real-time computations. Therefore, the presented mathematical model has a practical component rather than an academic value solely. In addition, the results were supported with extensive computer simulations. The proper choice of the control methodology is crucial to achieve the accurate needle placement under different scenarios such as the variety of tissue responses to the insertion force, different lateral components of the forces to the needle and, consequently, the needle deflection. Ideally, the control method should be able to correct the needle deflection during the insertion, or to prevent the deflection by minimizing the lateral insertion forces. In the proposed method, we apply the model predictive control (MPC) as a possible control method capable of achieving the desired dynamical behaviour of the system. The detailed implementation of the MPC prior to the simulation of the system was also presented in this article.

2 Mathematical Model of the System

To be able to perform the simulation of the system and to analyse the needle insertion parameters, the following steps were performed: (a) analysis of kinematics of the PIN, (b) development of the dynamic equations of motion for the PIN, and (c) development and simulation of a kinematic and dynamic control algorithm using the MPC.

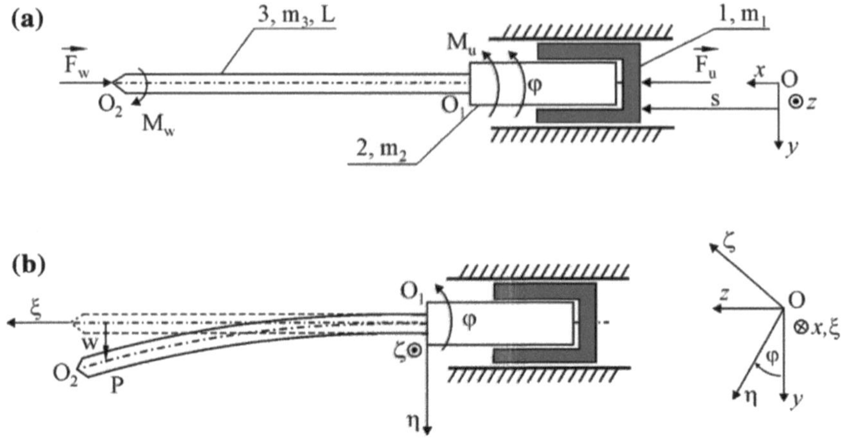

Fig. 1. (a) External schematic view of the mechanism with a needle (PIN), (b) The mechanism in the arbitrary moment during the needle insertion. The fixed and movable coordinate systems were presented.

Figure 1a represents the model of the proposed mechanism with a needle named PIN. It consists of the needle holder 1 having mass m_1. The needle holder moves in the translational direction along the straight guidance rail. The model of the mechanism included a rotational part of the needle holder, marked as 2, having mass m_2 and axial inertia moment I_2 with respect to the rotational axes. The elastic needle 3 is the third part of the mechanism connected to the end O_1, and it is represented as a cantilever beam. In Fig. 1a, s denotes the motion of the needle holder 1 along the guiding rail, and φ is the rotational angle of needle holder 2. The control force F_u acts upon the holder and control torque M_u acts upon the rotational joint 2. During the needle insertion procedures, the resistant torque M_w acts against the needle rotation as well as resistant force F_w that acts upon the needle tip O_2, as shown in Fig. 1a. It is assumed that the force F_w is parallel to the axis x during the motion of the needle.

The needle itself is represented as an elastic beam with n DOF. With this purpose in mind, the analyzed mechanical system has indefinite degrees of freedom. Since the mechanism consists of rigid and elastic bodies, the system belongs to the group of hybrid systems. The dynamical behavior of such systems is represented with both differential equations and partial differential equations. To increase the real-time calculations without losing precision, the methodology called the assumed-modes method was used for the mathematical modeling [3]. In this approach, the dynamics of the elastic needle was analyzed using a finite number of degrees of freedom. The configuration of the mechanism at the arbitrary moment during the needle insertion was presented in Fig. 1b. The frame O_{xyz} is the inertial frame of reference with O_{xy} representing a vertical plane. The moving coordinate frame $O_1\xi\eta\zeta$ is fixed to body 2 in such manner that the axis ξ coincides with the needle axis in its non-deformed configuration. The differential equations of the motion of the system considered can be obtained by means of the Lagrange equation,

$$\frac{d}{dt}\left(\frac{\partial L}{\partial \dot{\mathbf{q}}_r}\right) - \frac{\partial L}{\partial \mathbf{q}_r} = \mathbf{Q}_r, \; \frac{d}{dt}\left(\frac{\partial L}{\partial \dot{\mathbf{q}}_f}\right) - \frac{\partial L}{\partial \mathbf{q}_f} = \mathbf{Q}_f \tag{1}$$

where $L = T - V$ is the Lagrangian function. Equation (1) can be written in the following matrix form:

$$\begin{bmatrix} \mathbf{M}_{rr} & 0_{2\times n} \\ 0_{n\times 2} & \mathbf{M}_{ff} \end{bmatrix} \begin{bmatrix} \ddot{\mathbf{q}}_r \\ \ddot{\mathbf{q}}_f \end{bmatrix} + \mathbf{h}(\dot{\mathbf{q}}_r, \dot{\mathbf{q}}_f, \mathbf{q}_f) + \begin{bmatrix} 0_{2\times 2} & 0_{2\times n} \\ 0_{n\times 2} & \mathbf{K} \end{bmatrix} \begin{bmatrix} \mathbf{q}_r \\ \mathbf{q}_f \end{bmatrix} = \begin{bmatrix} \mathbf{Q}_r \\ \mathbf{Q}_f \end{bmatrix} \tag{2}$$

where, $\mathbf{h}(\dot{\mathbf{q}}_r, \dot{\mathbf{q}}_f, \mathbf{q}_f) = [0, 2m_3\left(\sum_{i=1}^{n} q_{if}\,\dot{q}_{if}\right)\dot{\varphi}, -m_3 q_{1f}\dot{\varphi}^2, -m_3 q_{2f}\dot{\varphi}^2, \ldots, -m_3 q_{nf}\dot{\varphi}^2]^T$
$R^{(2+n)\times 1}$.

This model represents the basis for the development of the needle insertion control that should minimize the lateral insertion force and needle deflection. The model does not incorporate the potential energy of the needle due to gravity force since it does not influence the overall system dynamics. Finally, the system (3) can be transformed to a non-linear state-space form as follows:

$$\begin{aligned} \dot{\mathbf{x}}(t) &= A(t)\mathbf{x}(t) + B(t)\mathbf{u}(t) + \mathbf{v}(t) \\ \mathbf{y}(t) &= C(t)\mathbf{x}(t) + \mathbf{w}(t) \end{aligned}, \tag{3}$$

where **x**(t), **u**(t) and **y**(t) are state vector, control vector and output vector, respectively; **v**(t) and **w**(t) are state disturbance and measured disturbance vectors; A(t), B(t) and C(t) are corresponding time varying state matrix, input matrix and output matrix, respectively. Following *the procedure reported in* [20], the matrices A(t), B(t) and C(t) are calculated as: A(t) = **[0 0 I 0; 0 0 0 I; 0 0 0 0; 0 K/M$_{ff}$ 0 0]**, B(t) = **[0 0 1/M$_{rr}$ 1/M$_{ff}$]**^{T}, and C(t) is equal to identity matrix since the generalized positions and velocities are chosen to be the system output. In addition, **x**(t) = $[\mathbf{x}_1 \mathbf{x}_2\ \mathbf{x}_3\ \mathbf{x}_4]^T$, $\mathbf{x}_1 = \mathbf{q}_r$, $\mathbf{x}_2 = \mathbf{q}_f$, $\mathbf{x}_3 = \dot{q}_r$, and $\mathbf{x}_4 = \dot{q}_f$. Consequently, **v**(t) = $[\mathbf{0\ 0\ 0} - \mathbf{h}_1(\mathbf{x}_2, \mathbf{x}_3, \mathbf{x}_4)]^T$, where $\mathbf{h}_1(\mathbf{x}_2, \mathbf{x}_3, \mathbf{x}_4)$ is the second element of the $\boldsymbol{h}(\dot{q}_r, \dot{q}_f, q_f) = [0, \boldsymbol{h}_1(\dot{q}_r, \dot{q}_f, q_f)]$ transformed to its state space form.

3 Model Predictive Controller

The MPC operates by finding the adequate control solution in an iterative process, incorporating the system dynamics, measurements and disturbances into the calculations. The schematic representation of the process is presented in Fig. 2. In this specific case, the formulation of the control procedure is as follows: based on the measurements of the resistance force F_w that acts upon the needle tip, the controller predicts the system behavior over the prediction horizon T_p, and calculates the input (F_u and M_u) of the system over the control horizon T_c. This calculation is performed such that the performance objective is minimized. In this case, the lateral displacement of the needle $\mathbf{R}_w$ is minimized.

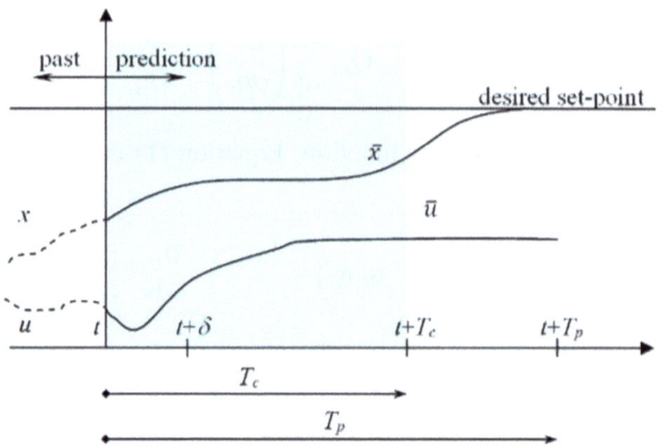

Fig. 2. General representation of the MPC functionality.

Consequently, the basic task of the MPC is to calculate the insertion speed represented by the motor driving force F_u, and the rotational speed of the needle represented by the motor driving torque M_u so that the lateral displacements of the needle tip and the lateral components of the insertion force over the sampling period can be minimized. This calculation is performed iteratively in real-time to allow for the needle steering with variable translational and rotational speeds. In the case when the lateral force components and the lateral displacement are zero, the controller continues to operate, using the same parameters until the desired behavior is maintained. The sampling time δ can vary to save the calculation time, but it is usually set to be fixed. The calculated system states at the instant $t + \delta$ let both the control procedure and the prediction horizon move forward. For this case, the linear model of the system would be insufficient to describe the process adequately; therefore, a nonlinear model was used for a more accurate representation of the process.

The overall purpose of the control task is to change the control parameters, such as the rotational and translational speed, and to obtain the rectilinear insertion makes the reference constant. In this work, we analyze a more conservative case – rectilinear needle insertion; however, this reference does not need to be constant and the system is potentially capable of needle steering in the 3D space. In general, to allow for the optimal needle placement inside the tissue for the brachytherapy procedures, we propose to minimize the lateral components of the insertion force (resistance force), and to predict and compensate for the influence of both the measured and unmeasured disturbances before their effect influences the system behavior i.e. needle displacement from the desired position. With this control strategy, it is possible to minimize both forces during the insertion and lateral displacement and, consequently, the needle deflection. The system receives information about the measured reactive force and tunes the parameters such as the translational speed and needle rotation in real-time during the insertion.

Referring to Fig. 3, the following parameters and variables are represented in the block diagram: D_{unm} represents the unmeasured disturbances to the system. These disturbances, such as the tissue resistance or lateral forces, increase the needle deflection during the procedure. The deflection depends on the insertion force, and based on that

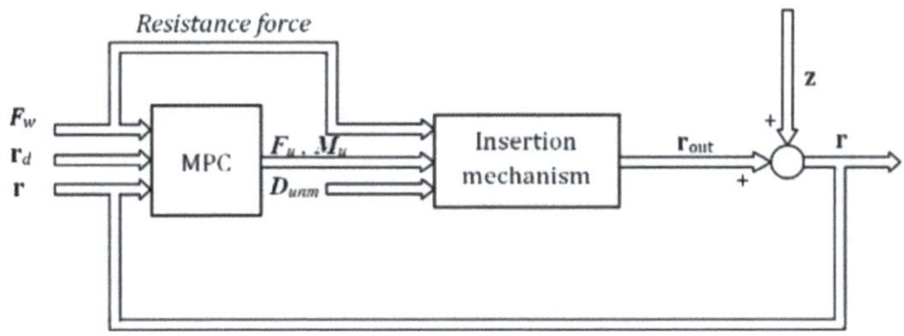

Fig. 3. Block diagram of the MPC for the needle insertion procedure using a proposed mechanism.

value, it can be estimated with a satisfactory level of accuracy, as in [1]. $\mathbf{r}_d$ is the vector that represents all reference position values, i.e. the desired needle tip position, during the insertion. This value is the targeting value of the output. The task of the system is to decrease the lateral components of the actual insertion force to minimize the deflection and, consequently, to meet the criteria for this setpoint. The resistance force F_w is a measured disturbance of the system for which the MPC provides a feedforward compensation to decrease its influence on the needle position. F_u and M_u are the generalized torques of the motors, as previously described. These control signals are adjusted in real-time to achieve the objectives of the system. The output of the systems is denoted as $\mathbf{r}_{out}$, and it is a real needle tip position during the insertion. This value is a controlled variable of the system. $\mathbf{r}$ is the output of the system, i.e. the calculated tip position that was used for the estimation of $\mathbf{r}_{out}$. The signal $\mathbf{z}$ represents the disturbance vector that contains the unpredicted influences on the system. The disturbance vector includes the electrical and actuator noises, calibration uncertainties, and inaccuracies due to the modeling of the physical system as well as other unpredicted effects on the system. The described approach is challenging and highly dependent on the accuracy of the mathematical model of the process. That is the basic reason why the equation of the motion (2) is crucial for further development. The second challenge in this approach is the proper individual tuning of the parameters, such as the adjustments of the actuators and the selection of the proper weights to allow for a minimal deviation of the desired deflection over its set-points within the prediction horizon.

In the situation when complicated physical processes related to needle tissue interaction are analyzed, two opposite requirements arouse: the possibility of the development of an accurate mathematical model and, on the other hand, the possibility to implement the proposed method to the real-time procedure with highly demanded data processing and computation. The insertion procedure should be performed only in few seconds. The most suitable controller for such tasks should can minimize the influence of the unknown system disturbances and deviations in mathematical modeling, while obtaining the measurements from the environment for the corrected control output. Therefore, it was decided to use a model predictive controller (MPCN). The MPCN is the key component of the proposed system. The proposed controller can predict and compensating for the unmeasured disturbances, such as the needle deflection or tissue reactive force, and it can correct them without waiting until the effect appears at the output of the system. On the other hand, the controller can maintain the desired needle tip position within the predefined limits, allowing for a high level of accuracy. Since the system model (3) is the system of a differential non-linear equation, the MPCN is also non-linear. It means that this approach required the iterative solution of an optimal control problem.

In this part, the basic functionality of the MPC is described. Initially, the MPC controller solves the linear-quadratic-Gaussian optimization problem for our multi-input multi-output system. For the system (25), the following cost function has been minimized,

$$J = E\left(\mathbf{x}^T(T)F\mathbf{x}(T) + \int_0^T (\mathbf{x}^T(t)Q(t)\mathbf{x}(T) + \mathbf{u}^T(t)R(t)\mathbf{u}(T))dt\right) \tag{4}$$

where E denotes the vector of the reference values or setpoints; T is the time horizon that can be finite or equal to $+\infty$, depending on the time during which it is necessary to perform the insertion of the individual needle. When $T \rightarrow +\infty$, $\mathbf{x}^T(T)F\mathbf{x}(T)0$. The goal of the optimization process is to find the feedback gain matrix $L(t)$ i.e. the proper torques for the motors that will guarantee the desired dynamic behavior of the system. Consequently,

$$L(t) = R^{-1}(t)B^T(t)S(t) \tag{5}$$

where $S(t)$ is the solution of the following Ricatti differential equation:

$$\dot{S}(t) = -A^T(t)S(t) - S(t)A(t) + S(t)B(t)R^{-1}(t)B^T(t)S(t) - Q(t) \tag{6}$$

The control algorithm with multiple variables uses the optimization function (30) together with the internal dynamics (25) of the system. For the control process, the information about the past controlled motions was used to calculate the optimal dynamical behavior during insertion.

$$J = \sum_{i=1}^{N} w_{x_i}(r_i - x_i)^2 + \sum_{i=1}^{N} w_{u_i}\Delta u_i^2 \tag{7}$$

where x_i is the needle tip position, r_i is the reference needle tip position, u_i is the control signal from the actuators, w_{x_i} is the weighting coefficient responsible for the relative importance of the x_i, whereas w_{u_i} is the weighting coefficient that penalizes the large change in gradient of u_i. Thus, the optimization cost function with the proper choice of the coefficients (7) ensures the decay of the transient errors as well as the reduction of the steady-state errors.

4 Results

The insertion accuracy was evaluated when the MPCN governed the needle insertion system represented by the Eq. (3). A representative case of the needle tip trajectory was presented in Fig. 4. Under the influence of the measured force, it was revealed that it was possible to minimize the needle deflection and to insert the needle into the desired position with the highest level of accuracy when the MPC approach was used. Additionally, Fig. 4 shows the simulation results when the $M_u = 0$ *i.e.* when the actuator dedicated to the needle rotation was turned off.

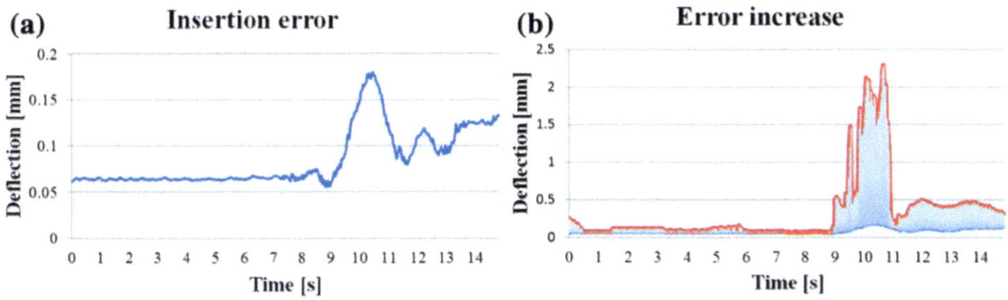

Fig. 4. Needle deflection (a) with and (b) without rotation control component (a representative case). An increase in the error was noticed when the needle rotation was not included into the process (marked region).

The simulation was performed under the influence of the recorded forces during the experiments. This conclusion proved the concept that the control of the translational motion was not sufficient and could not minimize the deflection up to the level at which the needles would not deflect significantly. Consequently, in this simulation, we proved the importance of the needle rotation for the accurate needle placement. When the simulation was repeated for different insertion depths (from 4 cm to 10 cm), it was revealed that the average needle tip displacements in the x, y and z directions were, −0.09 mm, 0.19 mm and 0.12 mm with an average SD = 0.13 mm, respectively. The auto-correction of the needle heading angle was presented in Fig. 5. It was observed that the angle increased initially before the MPCN started to correct the lateral displacement and, consequently, to minimize the angle.

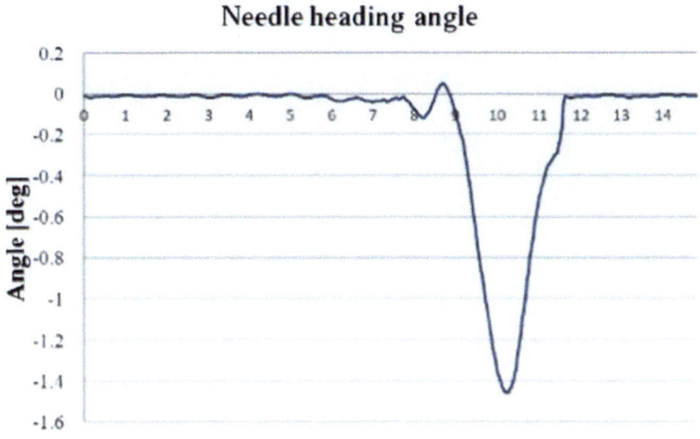

Fig. 5. Heading angle during insertion; MPCN corrects for the possible increase of the angle.

At this point, when the needle tip displacement is known, it is possible to use Eqs. (1) and (3) to recalculate the insertion forces. The values of the actuator parameters that lead to the insertions with minimal deflections are known as well. With the forward kinematics calculations of the system model, it is possible to calculate and re-evaluate the insertion forces achieved in the simulations. For this purpose, we recalculated the forces and the torques to evaluate if this process would result in any change of the measured insertion parameters. The results are presented in Fig. 6.

It was noticed that the torque on the insertion axis (z) did not change significantly whereas the x and y torques were minimized. This result was to be expected only if the lateral displacement of the needle was minimized. The torque on the insertion axis was not changed since it was necessary to apply the torque to allow for the needle insertion and steering. The similar observation was noticed in the case of the insertion force recorded at variable points of the needle during the procedures.

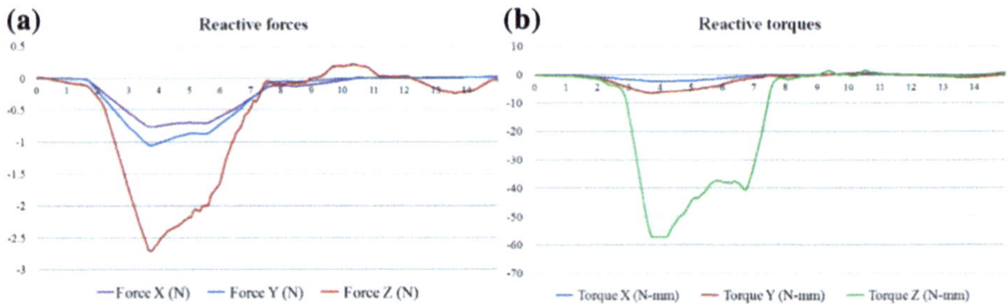

Fig. 6. Recalculated needle insertion parameters: (a) reactive forces in x, y and z directions, and (b) reactive torques on x, y and z axis (a representative case).

We compared the needle tip position accuracy to the accuracy achieved by the commonly used control strategies to evaluate the insertion accuracy when the MPC control approach and MPCN were used. The insertion system (Fig. 1) described by a mathematical model (2) was simulated using both the MPCN and the PID to evaluate the increase in accuracy when the MPC was used. The results were presented in Fig. 7.

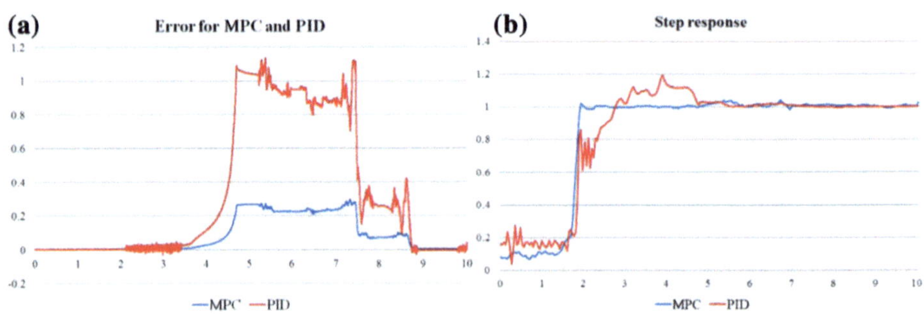

Fig. 7. Position error and (b) step response for MPC and PID needle control (a representative case).

It was noticed that the needle positioning accuracy was higher with the MPC method due to the needle steering provision, i.e. the adaptive change of the control parameters such as the velocity and rotational speed of the needle. When the response of both systems to the step function was evaluated, it was noticed that the transient parameters such as overshoot and settling time were more favorable in the MPC case. Whereas it is known that the PID ensures the asymptotic decay of the transient errors as well as reduction of the steady-state errors, the MPCN can deal with the overall system error without waiting until the effect appears at the output of the system regardless of whether the error is caused by the different modeling approaches or other unknown influences on the analyzed system. The overall conclusion was that the MPC approach was a preferred control methodology for the automatic needle insertion due to the described advantages over the PID control.

5 Conclusion

In this article, we presented a novel approach to the needle insertion. The controlled needle insertion using a predictive controller required a high level mathematical model of the physical process. Due to that fact, the mathematical model of the robotic system with (n + 2) DOF was presented. The needle was an elastic beam modelled with n DOF using the assumed-mode method. Using the adequate control methodology (MPC), it was possible to minimize the reactive force that was directly responsible for the needle deflection and needle tip displacement. Consequently, it was possible to guide the needle in a way which allowed for the maximum accuracy and minimal needle deflection. The future work will include the manufacturing of a mobile, portable and programmable robotic needle insertion mechanism with potential to be implemented in various medical procedures where the needle insertion must be performed with high reliability and accuracy, such as in interstitial brachytherapy, biopsy and other relevant procedures. The comparison between other control methodologies (such PID or adaptive control) and the MPC can potentially lead to clinically relevant results. The non-rectilinear needle insertion including the variable reference can be additionally investigated. In addition, the clinical trials using the proposed device are part of the future work so that the limitations and the possible advantages of this device should be adequately addressed in various clinical settings.

References

1. Buzurovic, I., Podder, T.K., Yu, Y.: Prediction control for brachytherapy robotic system. J. Robot. **1**, 1–10 (2010)
2. Buzurovic, I., et al.: Real-time control strategy for collision avoidance and seed deposition in brachytherapy robotic system. Int. J. Comput. Assist. Radiol. Surg. **3**, 30–34 (2008)
3. Siebert, F.A., Hirt, M., Niehoff, P., Kovács, P.: Imaging of implant needles for real-time HDR-brachytherapy prostate treatment using biplane ultrasound transducers. Med. Phys. **36** (8), 3406–3412 (2004)

4. Abolhassani, N., Patel, R., Moallem, M.: Trajectory generation for robotic needle insertion in soft tissue. In: Proceedings of the 26th IEEE International Conference on Engineering in Medicine and Biology (EMBS), San Francisco CA, USA, pp. 2730–2733 (2004)
5. Alterovitz, R., Goldberg, K., Pouliot, J., Taschereau, R., Hsu, I.-C.: Sensorless planning for medical needle insertion procedures. In: Proceedings of the IEEE/RSJ International Conference on Intelligent Robots and Systems, Las Vegas, NV, USA, pp. 3337–3343 (2003)
6. Chui, C.-K., Teoh, S.-H., Ong, C.-J., Anderson, J.H., Sakuma, I.: Integrative modeling of liver organ for simulation of flexible needle insertion. In: Proceedings of the 9th IEEE International Conference on Control, Automation, Robotics and Vision (ICARCV), Singapore, pp. 1–6 (2006)
7. Crouch, J.R., Schneider, C.M., Wainer, J., Okamura, A.M.: Velocity-dependent model for needle insertion in soft tissue. In: Proceedings of the 2005 Medical Image Computing and Computer-Assisted Intervention (MICCAI 2005), pp. 624–632 (2005)
8. DiMaio, S.P., Salcudean, S.E.: Needle insertion modeling and simulation. IEEE Trans. Robot. Autom. **19**(5), 864–875 (2003)
9. DiMaio, S.P., Salcudean, S.E.: Interactive simulation of needle insertion models. IEEE Trans. Biomed. Eng. **52**(7), 1167–1179 (2005)
10. Okamura, A.M., Simone, C., O'Leary, M.D.: Force modeling for needle insertion into soft tissue. IEEE Trans. Biomed. Eng. **51**(10), 1707–1716 (2004)
11. Wan, G., Wei, Z., Gardi, L., Downey, D.B., Downey, D.B., Fenster, A.: Brachytherapy needle deflection evaluation and correction. Med. Phys. **32**(4), 902–909 (2005)
12. Buzurovic, I., Tarun, K., Yu, Y.: Robotic systems for radiation therapy. In: Dutta, A. (ed.) Robotic Systems: Applications, Control and Programming, pp. 85–106. InTech, Rijeka (2012)
13. Fichtinger, G., Burdette, E.C., Tanacs, A., Patriciu, A., Mazilu, D., Whitcomb, L.L., et al.: Robotically assisted prostate brachytherapy with transrectal ultrasound guidance: phantom experiments. Brachytherapy **5**, 14–26 (2006)
14. Lin, A., Trejos, A.L., Patel, R.V., Malthaner, R.A.: Robot-assisted minimally invasive brachytherapy for lung cancer. Telesurgery **4**, 35–52 (2008)
15. Meltsner, M.A., Ferrier, N.J., Thomadsen, B.R.: Observations on rotating needle insertions using a brachytherapy robot. Phys. Med. Biol. **52**, 6027–6037 (2007)
16. Moerland, M.A., Van den Bosch, M.R., Lagerburg, V., Battermann, J.J., Van Vulpen, M., Lagendijk, J.J.W.: An MRI scanner compatible implant robot for prostate brachytherapy. Brachytherapy **7**(2), 100 (2008)
17. Wei, Z., Wan, G., Gardi, L., Mills, G., Downey, D., Fenster, A.: Robot-assisted 3D-TRUS guided prostate brachytherapy: system integration and validation. Med. Phys. **31**, 539–548 (2004)
18. Buzurovic, I., Podder, T.K., Yan, K., Hu, Y., Valicenti, R., Dicker, A., Yu, Y.: Parameter optimization for brachytherapy robotic needle insertion and seed deposition. Med. Phys. **35** (6), 2865 (2008)
19. Buzurovic, I., Podder, T.K., Yu, Y.: Force prediction and tracking for image-guided robotic system using neural network approach. In: Proceedings of the IEEE Biomedical Circuits and Systems Conference (BioCAS), Baltimore MA, USA, pp. 41–44 (2008)
20. Buzurovic, I., Debeljkovic, D.: A Geometric approach to the investigation of the dynamics of constrained robotic systems. In: Proceedings of the IEEE International Symposium on Intelligent Systems and Informatics (SYSI), Subotica, Serbia, pp. 133–138 (2010)

Author Index

N. Mitrovic et al. (Eds.): CNNTech 2018, LNNS 54, pp. 231–232, 2019.
https://doi.org/10.1007/978-3-319-99620-2

MIX
Papier aus verantwortungsvollen Quellen
Paper from responsible sources
FSC® C105338

If you have any concerns about our products,
you can contact us on
ProductSafety@springernature.com

In case Publisher is established outside the EU,
the EU authorized representative is:
Springer Nature Customer Service Center GmbH
Europaplatz 3, 69115 Heidelberg, Germany

Printed by Libri Plureos GmbH
in Hamburg, Germany